Mines of the Alder Creek Mining District of Idaho

by Idaho Geological Survey

with an introduction by Kerby Jackson

Introduction

It has been years since the Idaho Bureau of Mines released his important publication "History of Selected Mines of the Alder Creek Mining District". First released in 1997, this important volume has now been out of print for this days and has been unavailable to the mining community since those days, with the exception of expensive original collector's copies and poorly produced digital editions.

It has often been said that "*gold is where you find it*", but even beginning prospectors understand that their chances for finding something of value in the earth or in the streams of the Golden West are dramatically increased by going back to those places where gold and other minerals were once mined by our forerunners. Despite this, much of the contemporary information on local mining history that is currently available is mostly a result of mere local folklore and persistent rumors of major strikes, the details and facts of which, have long been distorted. Long gone are the old timers and with them, the days of first hand knowledge of the mines of the area and how they operated. Also long gone are most of their notes, their assay reports, their mine maps and personal scrapbooks, along with most of the surveys and reports that were performed for them by private and government geologists. Even published books such as this one are often retired to the local landfill or backyard burn pile by the descendents of those old timers and disappear at an alarming rate. Despite the fact that we live in the so-called "Information Age" where information is supposedly only the push of a button on a keyboard away, true insight into mining properties remains illusive and hard to come by, even to those of us who seek out this sort of information as if our lives depend upon it. Without this type of information readily available to the average independent miner, there is little hope that our metal mining industry will ever recover.

This important volume and others like it, are being presented in their entirety again, in the hope that the average prospector will no longer stumble through the overgrown hills and the tailing strewn creeks without being well informed enough to have a chance to succeed at his ventures.

Kerby Jackson
Josephine County, Oregon
October 2015

History of Selected Mines in the Alder Creek Mining District, Custer County, Idaho

Victoria E. Mitchell[1]

INTRODUCTION

The Alder Creek mining district is on the eastern slopes of the White Knob Mountains near the southern border of Custer County (Figures 1 and 2). Most of the mines are located near the contact between the White Knob Limestone and the Mackay stock (Figure 3). Ore was first discovered in the area in 1879, but the original prospectors were too poor to develop their claims. In 1884, additional discoveries, including one in the area of what would become the Darlington Shaft of the Empire Mine, finally set off a boom in the area. A 50-ton smelter was built at Cliff City on Cliff Creek, not far from the southern end of the Empire property (Figure 2; Umpleby, 1917). It began operations on November 23, and the initial test run produced a pure-looking product. However, the smelter was closed in early December after operating little more than a week (Wells, 1983). The smelter made another short run in 1885 but was again idle by January 1886. After Wayne Darlington gained control of the Empire, he persuaded some New York investors to help finance the mine. They supported work in the district in 1890, and the Cliff City smelter ran from late 1890 to February 1891 (Wells, 1983). British investors financed

[1]Idaho Geological Survey, Main Office at Moscow, University of Idaho, Moscow.

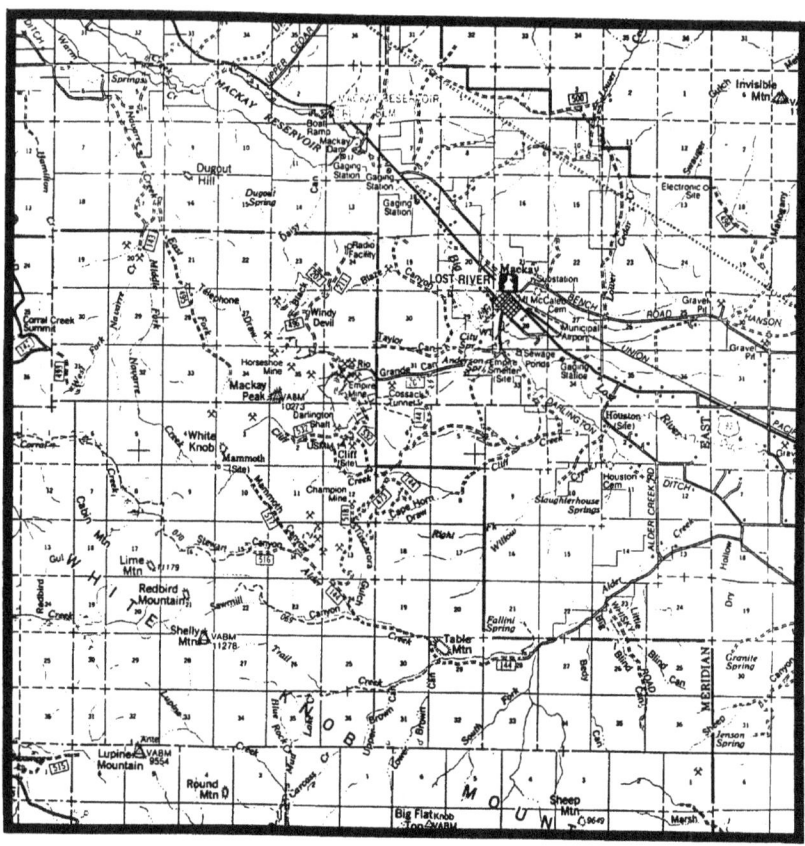

Figure 1. Location of the Alder Creek mining district and vicinity, Custer County, Idaho (U.S. Forest Service Challis National Forest map, scale 3/8 inch = 1 mile).

2

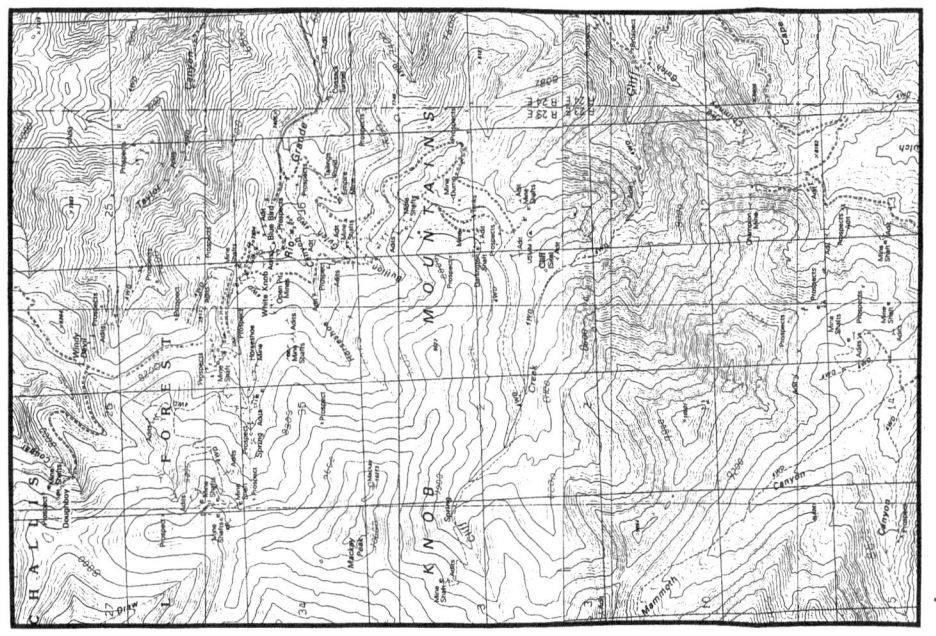

Figure 2. Topographic map of the area surrounding the Empire Mine, Alder Creek mining district, Custer County, Idaho (U.S. Geological Survey Mackay Reservoir and Shelley Mountain 7.5-minute topographic maps). Note the extensive mine workings and prospect locations throughout the area.

the construction of a good road to "a large Alder Creek mine" in 1892, but withdrew before doing any significant development at the mine. In 1894, W.A. Clark (whose holdings dominated the Butte district) began an extensive exploration and testing program (Wells, 1983).

Although Umpleby (1917, p. 100) claimed "The history of the district is essentially the history of the mine of the Empire Copper Co.", there are numerous other mines and prospects in the Alder Creek district (Figure 2). According to Nelson and Ross (1968), nearly 50 properties have been worked at one time or another, and the entire district has been heavily prospected. The Empire was by far the largest producer in the district, but other mines with significant production include the Homestake, Horseshoe, Doughboy, Blue Bird, and Champion mines.

Early production records for the district are not available, and how much metal was produced by the early smelter runs is not known. According to Wells (1983), $8,000 worth of copper was produced in 1899, presumably from the Empire Mine. Nelson and Ross (1968) stated that some ore was probably mined from the Empire Mine before 1902, but no records exist. Leland (1957) also believed the Empire produced high-grade oxidized ores from open cuts starting around 1884. Between 1902 and 1979, the six largest mines in the district produced a combined total of 994,269 tons of ore. From this was obtained 41,997 ounces of gold, 1,774,889 ounces of silver, 62,234,080 pounds of copper, 15,101,855 pounds of lead, and 5,496,067 pounds of zinc (Table 1).

Table 1. Total production from the largest mines in the Alder Creek mining district.

Mine	Ore (tons)	Gold (ounces)	Silver (ounces)	Copper (pounds)	Lead (pounds)	Zinc (pounds)
Blue Bird Mine (1918-1939)	1,530	23.44	16,426	12,595	509,165	3,111
Champion Mine (1908-1964)	2,281	14.20	7,236	20,003	579,668	32,719
Doughboy Mine (1919-1968)	1,070	8.30	31,702	1,972	637,120	3,900
Empire Mine (1902-1975)	921,077	41,431.25	1,294,531	61,689,291	24,110	906,078
Horseshoe Mine (1916-1979)	16,810	110.89	129,686	257,945	3,896,442	1,113,821
White Knob Mine (1909-1968)	51,501	408.53	295,308	252,274	9,455,350	3,436,438
TOTAL	**994,269**	**41,996.61**	**1,774,889**	**62,234,080**	**15,101,855**	**5,496,067**

EMPIRE MINE

The Empire Mine (Figures 2 and 3) was located around 1895. The mine was originally known as the White Knob, after the company operating it. (Table 2 lists the companies operating at the mine). A later White Knob Mining Company operated a different "White Knob Mine" about a mile north of the Empire in the 1920s and afterward.

The mine is in a contact metamorphic deposit, and mineralization was often related to blocks of White Knob Limestone engulfed by the Mackay stock (Figure 3). The orebodies were highly irregular, varying greatly in both size and shape, with branching arms that commonly diverged upwards and sometimes reconverged at higher levels. Most of the orebodies were circular or elliptical and most pitched to the southeast, but the exceptions were, in Umpleby's (1917, p. 46) words, "numerous and striking." The greatest vertical extent of an orebody described was from just above the No. 300 tunnel downward to the 850-foot level. In places this shoot was only a few feet wide, but in others, it widened to a floor area of up to 3,000 square feet (Umpleby, 1917).

In 1899, the main shaft at the White Knob Mine was 700 feet deep, with drifts connected by a ventilating shaft. Most of the ore was extracted from overhand stopes or by open quarry work. The cost of transporting supplies to the mine was 60 cents per hundredweight, and it cost $10 per ton to ship ore to the railroad. The average cost of smelting a ton of ore was also $10 per ton. With costs like these, large investments were needed to make any mine in the area successful.

Major financing for the Empire finally came from California millionaire John W. Mackay, who had made his money from the "Big Bonanza" (the Virginia Consolidated Mine in the Comstock Lode at Virginia City, Nevada). In 1901, Mackay arranged to have a branch of the Oregon Short Line constructed from Blackfoot to the Alder Creek district, providing his mine with ready access to rail transport. That same year, the town of Mackay, four miles northeast of the mine, was founded and soon had a population of 1,200. (Both the town and the Mackay School of Mines in Reno, Nevada, were named after Mackay.)

Development at the mine showed encouraging results. Exploration work had located a million tons of ore containing 4 percent copper, valued at $3 million in gold and copper (Wells, 1983). Wayne Darlington conducted a series of tests with the 50-ton smelter to verify that the ore could be processed without "insurmountable difficulties." The test lots produced 200,000 pounds of copper by direct smelting with no preliminary milling (Wells, 1983). Construction started on a 600 ton-per-day (tpd) smelter. The 1901 IMIR noted that the company employed "upwards of 500 men," was completing a 500-tpd smelter (the size of the smelter varied in different reports), and had constructed 12 miles of electric railroad to connect the mines to the smelter.

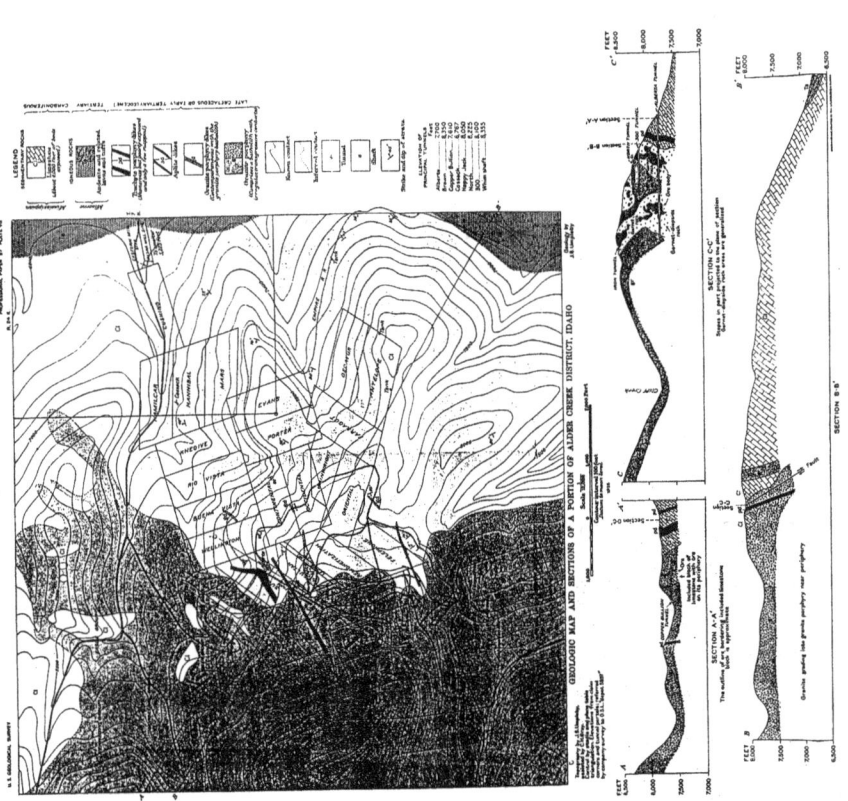

Figure 3. Geologic map and sections of a portion of the Alder Creek mining district. Map also shows claims and workings of the Empire Mine and the locations of other claim groups in the vicinity (Plate VII from Umpleby, 1917).

6

Table 2. Companies operating at the Empire Mine.

Company Name	Officer	Date Incorporated	Charter Forfeited	Year(s) at Mine
White Knob Copper Co.	John W. Mackay, President	[1]	[1]	[1]
MacBeth Lease, Inc.	Ravenal Macbeth	[1]	[1]	1904-1907
White Knob Copper and Development Co.	[1]	[1]	[1]	?-1905
Empire Copper Co.	Frank M. Leland, President	June 28, 1907	Dec. 1, 1921 (company reorganized as Idaho Metals Co.)	1907-1921
Idaho Metals Co.	L.R. Eccles, President	Oct. 8, 1921	[1]	1921-1928
(in receivership)	---	---	---	1928
Mackay Metals, Inc.	W.E. Narkaus, Manager	June 4, 1928	December 1, 1930	1928-1931
(in receivership)	J. Ray Weber, Receiver	---	---	1931-1936
Mackay Exploration Co.	Ted Cherry, President; J. Ray Weber, Manager	August 21, 1939; reinstated March 25, 1974	1971; 1974	1939-1960
Custer Copper Corp. (lessee)	W.P. Barton, President	June 28, 1946	active through 1967	1946-1956?
Idaho Alta Metals Corp. (lessee)	E.G. Bowen, Executive Vice President	November 19, 1954; reinstated January 24, 1957	1956?; November 30, 1959	1956-1958
R.V. Lloyd & Co.	R.V. Lloyd, President	[1]	company reorganized as Lost River Mines, Inc.	1960-1965
Lost River Mines, Inc. (Empire Copper, Inc.)	R.V. Lloyd, President	March 4, 1965	November 30, 1966	1965-1966
J.R. Simplot Co.	J.R. Simplot, President	February 2, 1946	active	1970
Ivie Mining Co.	W.W. Ivie, President	December 10, 1969; reinstated January 30, 1974	1971; 1975 (company taken over by Honolulu Copper Co.	1971-1974
Honolulu Copper Co.	[1]	[1]	[1]	1972-[1]
Myko, Inc.	Ivan Taylor, Vice President	March 7, 1973	not reported as active in 1981	1973/1974-[1]
Exxon	[1]	[1]	still active	exploration: 1977

[1]Information not available in IGS's files.

The following year, the company drove a long cross-cut tunnel from the head of the electric railway. Plans called for starting up the smelter after this tunnel reached the lowest workings of the mine.

Labor difficulties marked the mine's operations during 1902. Darlington was unable to get along with his miners. On April 4, 1902, they organized as a union in the Western Federation of Miners. Two days later, the union went on strike to avoid being driven out of the district. In the settlement of this dispute, Darlington, his superintendent, and his foreman were dismissed (Wells, 1983). John W. Mackay died in London on July 20, 1902. When one of the two furnaces for the smelter was completed in October 1902, none of the people originally responsible for the big smelter was still connected with the mine.

The 1903 IMIR contained the following description of the mine and its operations (p. 53-54):

> The White Knob mine is situated on a spur from White Knob peak at an elevation of eight thousand feet above sea level and two thousand feet above the valley of Lost River and the company's big smelter site. The surface manifestations of ore at this property amount to a mineral farm that cover the flat top of the ridge over an area of fully forty acres with low grade copper ore and copper stained formation in great patches and zones. Near a contact of a wide body of eruptive granite porphyry and overlying limestone beds, and including great masses of solid fifty to sixty per cent hematite and magnetite iron ore, which also carries important values in gold and silver.
>
> The development consists of a vertical shaft seven hundred feet deep sunk from the croppings[1]. This, however, is only used as an air shaft at present, for it has been supplanted with a cross-cut tunnel eleven hundred feet long that taps the ore on a level with the bottom of the shaft.
>
> The ore bodies occur in the blue limestone, also in contact with a large dike of included coarsely crystalline feldspar porphyry. The ore seems to be a replacement of the lime and occurs in mammoth shoots thirty to fifty feet wide and seventy-five to one hundred feet long. The ground is dry and the ore still mostly altered reddish brown oxides and green carbonate of copper carrying from two to four per cent of the red metal together with about two dollars in gold and silver [Figure 4].
>
> The main ore shoots have been persistently continuous from the surface down to the first big cross-cut called the Albert tunnel[2], with stations and drifts at convenient horizon. The work is very substantially timbered and equipped with adjustable chutes arranged for handling ore, iron or lime flux, all of which can be mined right on the ground. One of the largest ore bodies is showing sulphide mineral strong in stockwerk, threads, pebbles and masses of pure ore that varies from brassy chalcopyrite to spots of rich soft blue black bornite ore. This sulphide is very desirable for matting and has been eagerly anticipated by the management [Figure 5]. The strike of these great bodies is east

[1] The Darlington Shaft.

[2] On the 700 level. Later authors referred to this as the Alberta tunnel.

Figure 4. Typical specimens of ore from the Empire Mine (Plate XV from Umpleby, 1917). 'A' is chrysocolla ore, which varied widely in color and in content of copper, iron, and manganese. 'B' is carbonate copper ore, composed of malachite, azurite, chrysocolla, brochantite (hydrous copper sulphate), and banded amorphous silica.

Figure 5. Specimens of ore from the Empire Mine (Plate XVI from Umpleby, 1917). 'A' is closely associated malachite (light) and brochantite with veinlets of gypsum (Copper Bullion tunnel). 'B' is garnet and chalcopyrite (light specks) cut by a veinlet of gypsum (Copper Bullion tunnel). 'C' is cuprite (light) surrounded by tenorite (dark layer) in chrysocolla ore (300-foot level). 'D' shows covellite pseudomorphs after chalcopyrite (black areas and lines) from the 300-foot level. 'D'-'b' is magnified slightly less than 30X; all others are about 90 percent of natural size.

10

and west [of north][3] with a steep dip to the north [south][3] and towards the valley. A second tunnel has been started down the mountain side which will tap the ore bodies nine hundred feet below the Albert tunnel and sixteen hundred feet from the surface[4].

This new tunnel is making rapid progress and should reach the big ore bodies in the early summer. It is started at the level of a deep gulch that furrows the mountain side where more water and much higher copper values may be safely expected when the ore bodies are reached.

The White Knob property is equipped with one of the finest smelting plants in the west [Figure 6]. It has two large blast furnaces of three hundred fifty tons capacity each, with all the necessary attendant equipment, including an electric railway seven miles long connecting the smelter with the mine. The ore is smelted direct as it comes from the mine without preliminary milling. One of these furnaces was gotten into commission in October and with the slight delays incident to tightening the harness of a plant of this kind, has been in successful operation since, producing a high grade matt, which carries fifty-eight per cent copper and important values in gold and silver, with a slight loss of less than half of one per cent copper. As soon as the development of the mine is a little further along so that the necessary large daily tonnage can be handled economically the other furnace will be started and will give the plant its maximum daily capacity of seven hundred tons a day. The ore bodies already developed in the mine are said to contain a net value of something like $2,000,000 over the cost of their extraction and reduction, and with the advantage of their precious values and the definite prospect of increased copper percentage at the new level now being opened the chances are that the White Knob will become one of the largest, most permanent and profitable producers of copper, gold and silver bullion in the west.

Umpleby's (1917, p. 93) description of the smelter's initial operation was less optimistic:

"The endeavor was to make black copper, but the furnace was so proportioned that draft was almost impossible, and after running 30 days on 2 per cent ore and failing to reduce slag content below 1.4 per cent copper, matting was adopted. For this purpose sulphide ore was shipped from Butte, Mont., but the financial losses were great and at the end of 10 months operations ceased."

The White Knob Copper Company worked the mine actively for the first eight months of 1904, employing between 300 and 400 men. However, the company failed in September and the property passed into the hands of a receiver. According to the Idaho Inspector of Mines, the company was being reorganized and was soon expected to resume developing the mine. He went on to say (1904 IMIR, p. 53-54):

---

[3]Strike-outs and interpolated corrections are hand-written in IGS's copy of the 1903 IMIR. Changes agree with information given by Umpleby in U.S. Geological Survey Professional Paper 97 for ore bodies above the Alberta (Albert) tunnel level.

[4]The 1600 level tunnel was variously known as the Van-Austin, the Van Ostrand, the Cassock, the Cossak, and the Cossack.

Figure 6. The White Knob (Empire) smelter at Mackay, Idaho (preceding page 55 *in* Bell, Robert N., 1905. Report of the Mining Districts of Idaho for the Year 1904).

The great trouble with the White Knob enterprise is the fact that its expensive smelting plant and surface equipment were undertaken before the metallurgical feature and method of treating the ore was sufficiently worked out. The great bodies of carbonate and oxide ores have continued down 700 feet in that altered condition to the Albert Tunnel level, and experience has shown that the ore lacked sufficient sulphur for a successful matting method of treatment with which it has been attempted to work it, and too much sulphur for making base bullion successfully.

Both the sulphur and copper tenor of the ore has shown a marked increase in two short winzes that have been sunk from the Albert Tunnel, and if the development is followed out in this direction and subsequently connected with the Van-Austin Tunnel that is designed to tap the ore bodies 900 feet below the Albert Tunnel, and is already in over 1,000 feet, the property is likely yet to blossom ont [out] in all the glory promised by its early advocates, for it has immense bodies of low-grade, oxidized and carbonate ores, and if they follow the normal rule of big copper ore bodies, a zone of secondary enrichment is yet due to make its appearance in the downward extension of the ore bodies that should bring the property into favorable prominence, and eventually justify the splendid plant with which it is equipped.

Umpleby (1917) was more direct when discussing the operations at the mine between 1901 and 1905 (p. 13-14):

The succession of White Knob companies which owned this property during the next five years[5] is notorious in the annals of mining, each being a drain on the investing public and a failure more disastrous than the one preceding it. After an expenditure of about $3,000,000 without a cent of profit the enterprise passed into the hands of the Empire Copper Co., an entirely new organization, which has operated the mine on a leasing system at a noteworthy profit. The deceit and mismanagement that characterized its early history have been a serious detriment to the development of the mineral resources of the region, but its present management is conservative, and the company is encouraging the local industry in every legitimate way.

He expanded on this theme on p. 93:

During the period from about 1900 to 1907 the leading property of the district was owned and operated by a succession of White Knob corporations, which, though under the same general management, succeeded each other at short intervals, each a failure more disastrous than the one preceding it. Prior to 1905 more than $3,000,000 is said to have been spent on the property, and at one time the stock sold for $23 a share on a capitalization of 600,000 shares, yet not until 1905, when operated by a leasing company, was the mine ever worked on a business basis or at a profit.

The 1905 IMIR discussed the changes in management as follows (p. 52-53):

The White Knob Copper Mine, at Mackay, experienced one of the most interesting years of its checkered career during 1905.

---

[5]i.e., between 1901 and 1905.

13

This property was originally developed by Mr. Wayne Darlington, ex-State Engineer of Idaho, who did some extensive exploration work on the immense bodies of low-grade ore that the property carries and opened up a big resource of copper carbonate and oxide mineral near the surface. He built a 50-ton smelter and made several test runs of the ore from different parts of the property, producing a total of something like 200,000 pounds of base copper bullion by direct smelting.

This demonstration of handling the low-grade ore of this mine was so successful that its extensive equipment with a 600-ton smelter was undertaken and about half completed when one of the principal backers[6] of the enterprise died and the control of the property passed into the hands of others who were antagonistic to Mr. Darlington's plans. This resulted in the management being turned over to other people who changed his plan and design of the big smelter, transforming it into a matte plant, which in the hands of a number of high-priced operators, proved unsuccessful from lack of sufficient sulphur in the ore to make matte and save the values, with the result that last spring the property was turned over to a practical operator from California, Mr. Frank Leland, who was instructed to junk the plant and wind up the affairs of the company.

Mr. Leland found several thousand tons of low-grade ore on hand, and thought he could make a success of smelting it. He sold off all the superfluous supplies and equipments, put the company's assayer in charge of one of the furnaces at the smelter and put the mine into the hands of some intelligent leasors, who went to mining on the best ore and sweetened up the values of the mineral he had on hand, with the result that one of the big furnaces was kept running for several months and successfully handled the material on hand, producing twenty-five carload shipments of high-grade gold and silver-bearing copper matte, and slag result containing considerably less than one-half of 1 per cent copper. The smelter was shut down late in the fall and the leasing system at the mine extended.

Under Mr. Leland's leasing system the mine is now furnishing employment for about one hundred men, and recent reports show that they are putting out ore of a far better grade than was ever before produced from the mine, and the prospect seems bright at this date for the White Knob to become a profitable source of copper bullion after all its vicissitudes.

The White Knob ore deposits occur at a contact between limestone and porphyritic granite, or rather between an immense body of garnet rock, mixed with other contact metamorphic minerals and the overlying limestone.

The ore occurs in big, irregular shaped bodies 50 feet wide by 100 feet in length in some instances. It has continued in a largely oxidized and carbonated condition to a depth of 700 feet below the apex, where it has been tapped by a cross-cut tunnel.

The ore bodies seem to favor the limestone rather than the garnet rock. The deposit is distinctly of the Arizona variety, and the extensive leasing system now employed in the exploration of its ore bodies is not unlikely to lead to very valuable deposits of massive sulphide ore at further depth, as the ore is commencing to show some very handsome bodies of chalcopyrite mineral in the lowest opening.

The mineralization of this property is very extensive and justifies considerable further development. Mr. Leland has replaced the expensive electric haulage system [Figure 7] between the mine and smelter with a Shay engine [Figure 8], which it is expected will handle the material off the mountain at one-fourth of the former cost.

In 1904, Ravenal Macbeth (who had formerly worked the Lucky Boy at Custer and who would later be the Secretary of the Idaho Mining Association) obtained a

---

[6]Multi-millionaire John W. Mackay.

Figure 7. Loading bin and electric tram at the White Knob (Empire) Mine, 1904 (opposite page 54 *in* Bell, Robert N., 1905, Report of the Mining Districts of Idaho for the Year 1904).

Figure 8. Shay locomotive and ore cars at the White Knob (Empire) Mine, 1906 (opposite page 52 *in* Bell, Robert N., 1907, Eighth Annual Report of the Mining Industry of Idaho for the Year 1906).

lease at the White Knob (Wells, 1983). In November 1905, MacBeth Lease, Inc., secured a 5-year lease on the mine and smelter and immediately let a number of subleases (Umpleby, 1917). Most of the 1905 production came from the Macbeth operation. The smelter ran for several months, but at the end of the year, it was undergoing repairs and being adapted to process the ore produced by the lessees.

The Macbeth lease was the most important mining operation in Custer County during 1906. One hundred men worked the mine between the time the smelter closed in the fall of 1905 and March of the following year, developing "some extensive bodies of good copper ore," mainly near the surface outcrops of the deposit (1906 IMIR, p. 52). The Mine Inspector (1906, p. 53) noted that the orebodies occurred "in very irregular manner in pockets, kidneys, streaks and lenses over the area described and are mined in big open cuts and quarries." During the year, a fire destroyed the entire sampler, the machine shop, and the Shay locomotive, which caused a financial loss to the lessees of nearly $100,000. In spite of this, they produced 50,000 tons of ore during the year, which averaged over 4 percent copper, with significant amounts of gold and silver. One of the smelter's furnaces was blown in on March 5, 1906, and it operated continuously until August 29 of the following year. One factor in the smelter closure was complaints from ranchers against the smoke and from the fishermen against the slag dumped into the river (Umpleby, 1917). About 200 tons of ore per day was treated at the smelter.

Macbeth made substantial changes in the operation of the mine and plant (1906 IMIR, p. 53-54):

> The surface deposits on the property were quite extensively developed under the original manager's administration, an electric railway was built right up to the quarries, and a steam shovel purchased for the economical handling of the ore in the surface work. The trackage and equipment of this electric railway above the Albert tunnel and the steam shovel were sold off for junk by the last administration on the White Knob prior to the present lease and an attempt was made to convey the surface ores down through a system of raises to the Albert tunnel level seven hundred feet below. This proved impractical, however, and the leasers installed a surface gravity tram during the early part of the summer for handling this most important source of their ore supply down to their big loading bins at the Albert tunnel level from where it was hauled by Shay geared locomotive to the smelter at Mackay, six miles below, over a six per cent grade.

The lessees had to ship in iron sulfides from Bingham, Utah, to mix with the ore to furnish enough sulphur for the smelter to operate properly. The resulting matte ran about 40 percent copper and from $30 to $50 per ton in gold and silver. Slag losses were said to be less than 0.25 percent when using a feed of ore containing 4 to 6 percent copper and using 13 percent coke fuel. Leland (1957) states that mismanagement under different companies between 1901 and 1907 resulted in an operating loss of $3 million for that period.

The Macbeth lease operated the mine until June 6, 1907. At that time, the property was taken over by the Empire Copper Company of New York City, with

17

Frank M. Leland as the president and general manager. The new company employed about 275 men at the mine and was the largest mining operation in the county. The smelter was operated for part of the year, producing a high-grade copper matte. However, the new company negotiated an arrangement with the railroad and the Bingham smelter to ship the ore directly to the smelter. This provided for a better profit margin because it did not require the company to pay for shipping fuel and sulfur to Mackay as well as shipping the matte to Bingham. The company smelter at Mackay was never reopened.

Even with the favorable transportation rates, a slump in copper prices in September 1907 made it unprofitable to ship the grades of ore that the mine was producing (about 60 to 80 pounds of copper and a few dollars in gold and silver per ton). The mine closed on October 1, although the Mine Inspector noted that large reserves of ore were in sight and that the mine would be able to operate at a reasonable profit with copper prices in the 17-18 cents per pound range. (The average price of copper the following year was 13.2 cents per pound. The price continued to drop for the next 3 years, reaching 12.5 cents per pound in 1911, before going back up.) Development on the property totaled 5,650 feet of workings. Wages paid to the workers by the Empire Copper Company were: miners, $3.50-$4.00 per day; laborers, $3.00 per day; blacksmiths, $4.00-$4.50 per day; and carpenters, $4.00-$5.00 per day.

Lessees carried out most of the work at the Empire during 1908 and 1909. In 1909, they produced 10 carloads of shipping ore during the year. The mine was kept in good repair and was ready to begin operations as soon as the price of copper went up. Lessees shipped a large tonnage of low-grade copper ore during 1910. Approximately 834,000 pounds of copper, 24,000 ounces of silver, and 250 ounces of gold were produced. Lessees also did 150 feet of sinking and discovered new orebodies, which had not been fully evaluated by the end of the year.

In 1911, the Empire was the largest producer in Custer County (Figure 9). From May to October, 155 cars of ore were shipped from the mine to Salt Lake. This ore ore assayed about 6 percent copper and had an average value of $19.22 per ton. Over 1 million pounds of copper was shipped during this period. Total shipments for the year from all leasing operations were 220 carloads, which provided a decent profit for both the company and the lessees. The Mine Inspector stated (1911, p. 40):

> The leasing operations have been invariably profitable, as the ground is divided up into a number of smaller operations, and successful results depend on the care with which the ore is followed, handled and kept clean, and the leasing methods seem to attain this result much more successfully than has any of the several attempts at handling the property on a large scale by company management. The persistent success of these leasing operations is a strong argument in favor of the permanency of the deposit, which has been studied by some very eminent geological authorities, and it is believed that with more extensive development at depth a better segregation of its numerous branching ore channels would be encountered and a more profitable mine result.

Figure 9. Alberta tunnel, or 700 foot, level of the Empire Mine (c. 1910 or 1915) (Idaho Historical Society photograph).

The mine operated the entire year in 1912. Production more than doubled, reaching 25,000 tons of ore which averaged about 6 percent copper, $1.50 in gold, and 3 ounces of silver per ton. Most of the ore shipped ran over 5 percent copper. More than 100 men, mostly lessees, worked the mine. The strength of the ore showings in many of the leases led the company to proceed with its plans for deeper development through the Van Austin (Cossack) tunnel (which had been started in 1903). The tunnel was about 1,000 feet long, and it was hoped it would intersect the Alberta orebodies 900 feet below the Alberta tunnel (Figures 10 and 11).

In 1913, the Empire was the second largest copper mine in the state. (Table 3 lists mine output and economic data.) A force of 150 men worked throughout the year. The mine shipped 55 to 60 cars of ore a month to the smelters near Salt Lake City. The ore averaged about 6 percent copper with $5 per ton in gold and silver. The Cossack tunnel was over 2,000 feet long and was expected to undercut the working areas of the mine in another 1,000 feet. Mine equipment included two drill compressors, ten air drills, one 8-horsepower gasoline hoist, two 10-horsepower air hoists, 7 miles of narrow gauge railroad, two 31-ton Shay locomotives, 30 cars with a capacity of 7 tons each, and "One 500 ton Smelter (not running)". Total development was 36,960 feet of workings. (Table 4 lists development work at the mine.) The company paid out a 10-cent dividend during the year, which came to $100,000.

Umpleby (1917) visited the district in 1912 and 1913. At that time, aggregate mine workings were between 20,000 and 25,000 feet. The Cossack tunnel was 1,900 feet long and, Umpleby estimated, some 2,000 feet short of reaching a point below the north Alberta ore shoot. The other principal groups of workings were the Darlington shaft, the Alberta tunnel, and the Copper Bullion tunnel. The Darlington shaft was no longer accessible. The Copper Bullion tunnel was about 1,600 feet long, with connecting raises, winzes, and laterals totaling another 800 feet. The Alberta tunnel was the most important in the mine (Figures 12, 13, and 14). The main adit was 2,800 feet long and connected to the Darlington shaft; the lateral tunnels off the main adit added another 3,000 feet to the workings. There were also nine other tunnels, each between 100 and 1,000 feet long. Equipment at the mine included three hoists and an 8-drill compressor. The property was equipped with steam, gasoline, water, and air power. The company's railroad, which ran from the mine to Mackay, climbed 2,000 feet over the course of a circuitous 7¾-mile route. The railroad was equipped with two Shay mountain-climbing locomotives and 38 cars.

Production for 1914 was about half that of 1913 due to unfavorable market conditions related to the start of World War I. The Empire's market was completely shut off during the summer. However, by fall the company had a full complement of lease operations in place and over 60 men were working the mine. The Cossack tunnel was between 3,000 and 4,000 feet long, and, while still 1,400 feet short of the developed orebodies, was expected to reach the main ore zone within another year.

Figure 10. Principal workings at the Empire Mine. a = North Tunnel level; b = 300-foot level; c = Alberta or 700-foot level (Plate IX-A from Umpleby, 1917).

Figure 11. Principal workings at the Empire Mine; area is north of the area shown in Figure 10. The Alberta level is in the foreground; areas marked "ls" are blocks of limestone enclosed in granite porphyry (Plate IX-B from Umpleby, 1917).

Table 3. Mine output and economic data for Empire Mine for selected years, 1913-1931[1].

Year	Tons of ore	Average value per ton	Total mining cost per ton	Transportation and treatment costs per ton	Gold recovered (ounces)	Silver recovered (ounces)	Copper recovered (pounds)	Gross returns
1913	34,168	$18.33	$4.00	$4.79	2,243	77,148	3,507,281	$584,375.67
1915	23,450	14.01	3.50	4.63	1,236	76,728	2,767,100	328,534.50
1916	71,078.68	16.58	3.50	5.05	3,491.13	139,689.49	5,471,292.40	1,178,451.97
1918	64,437.78	19.84	[2]	5.60	2,740.77	104,280.86	4,481,645	1,278,377.69
1921	14,700	16.56	[2]	6.83	2,469.60	32,648.70	88,641	142,179.97
1922	10,404	[3]	[2]	3.30	957.17	23,259.18	1,072,028.16	[4]
1923	15,184	[3]	[2]	3.03	1,731.29	28,151.14	1,614,467.65	[4]
1924	15,000	[3]	2.00	2.75	1,500	25,500	1,350,000	[4]
1931	27,507	[3]	2.77	1.50	731	22,102	1,097,780	154,720.67

[1] All numbers taken from company reports to the Idaho Mine Inspector. Reporting periods and accounting methods are not necessarily consistent between companies or between reports made by different individual for the same company. Every effort has been made to accommodate varying reporting practices, where possible, but final results are only as good as the original data.
[2] Mining costs not given.
[3] Gross value per ton not given.
[4] Gross returns not given.

23

Table 4. Employment, development work, and operating companies at the Empire Mine, by year.

Year	No. of Men employed	Tunnels (feet)	Sinking (feet)	Cross-cutting (feet)	Drifting (feet)	Raising (feet)	Operator
1913	60/60[1]	1,500[2]	800[3]	700[4]	---	---	Empire Copper Co.
1914	[5]	3,000[2]	400[3]	2,600[4]	---	---	Empire Copper Co.
1915	165	4,000[2]	1,000[3]	3,000[4]	---	---	Empire Copper Co.
1916	293	9,000[2]	1,000[3]	8,000[4]	---	---	Empire Copper Co.
1917	274	---	2,724[3]	6,134[4]	---	---	Empire Copper Co.
1921	30	3,100[2]	1,240[3]	1,860[4]	---	---	Empire Copper Co.
1924	45	1,307	146	---	---	1,093	Idaho Metals Co.
1925	77	610	---	---	---	---	Idaho Metals Co.
1927	[5]	200	---	---	---	---	Mackay Metals, Inc.
1928	70	1,335[2]	---	---	---	---	Mackay Metals, Inc.
1930	90	2,200	---	---	---	---	Mackay Metals, Inc.
1939	/45	400	---	---	---	---	Mackay Exploration Co.
1941	28[6]	---	28	---	---	---	Mackay Exploration Co.
1942	22	810	386	---	---	273	Mackay Exploration Co.
1943	11[6]	166	15	---	---	98	Mackay Exploration Co.
1945	35	205	---	---	---	110	Mackay Exploration Co.
1947	15	200	50	---	---	---	Custer Copper Corp.
1948	15	150	35	---	---	---	Custer Copper Corp.
1950	4	[7]	---	---	---	---	Custer Copper Corp.
1962	20	---	---	---	150	---	R. V. Lloyd & Co.

[1]First number given is number employed by company. Number following "/" is the company's estimate of the number of lessees and men employed by the lessees. Where no slash is used, company did not report distinction between employees and lessees or did not report number of lessees.
[2]Total figure for development work for the year.
[3]Combined figure for sinking and raising.
[4]Combined figure for cross-cutting and drifting.
[5]Number of men employed not stated.
[6]Combined figure for company employees and lessees.
[7]Development work for year confined to surface stripping.

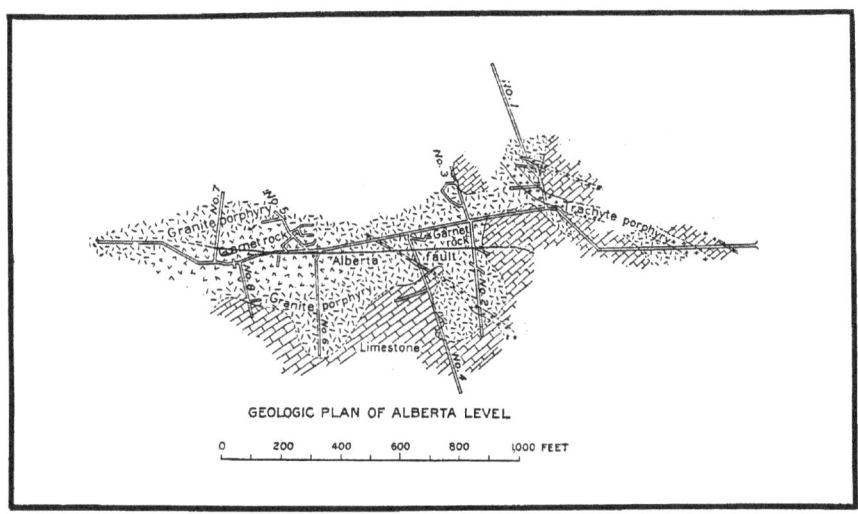

Figure 12. Generalized geologic plan of the Alberta level of the Empire Mine (upper half of Figure 2 from Umpleby, 1917).

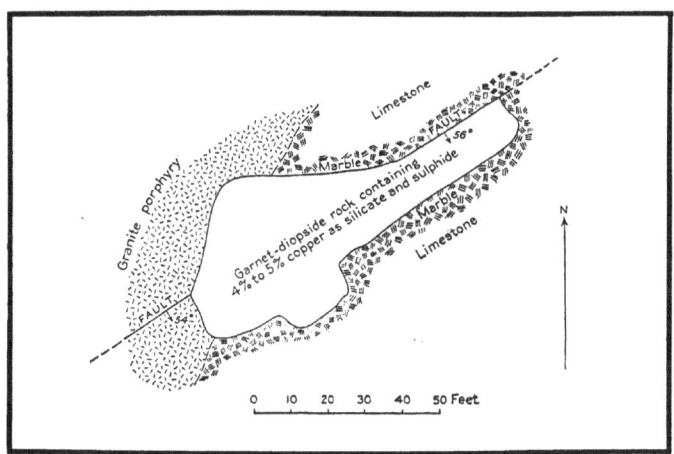

Figure 13. Generalized plan of the stope on the 450-foot level of the North Alberta ore shoot at the Empire Mine (Figure 3 from Umpleby, 1917).

25

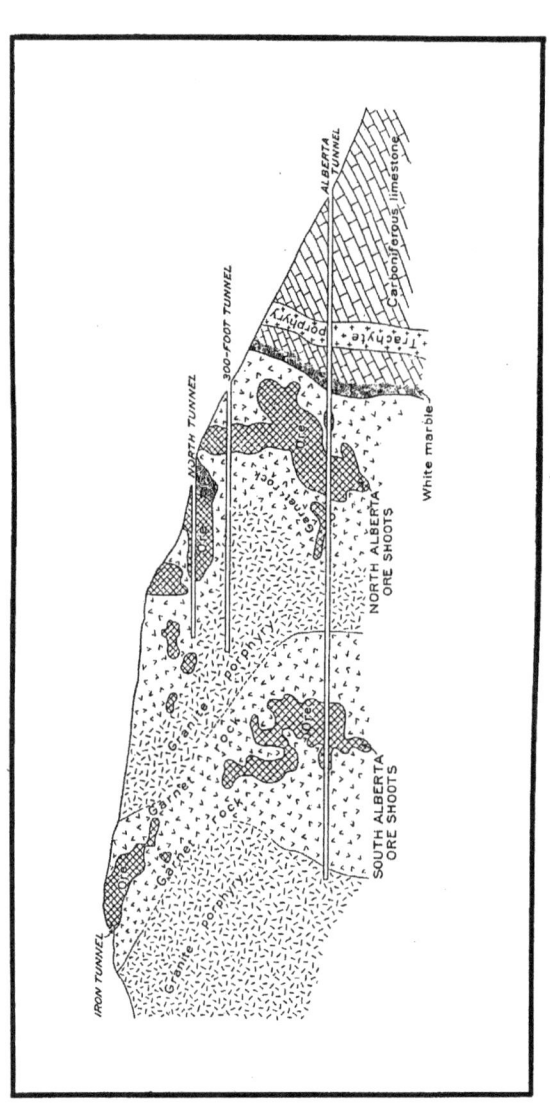

Figure 14. Longitudinal section through the Empire Mine, showing generalized geology (lower half of Figure 2 from Umpleby, 1917).

The Empire operated all year in 1915 and made particular efforts toward the end of the year to cash in on the increase in copper prices. This resulted in the company's most prosperous year ever. In each of several months, lessees shipped as much as 9,000 tons, but the average for the year was less than 5,000 tons a month. More ore could have been shipped if the smelters could have handled it. About 200 men were working the mine, most of them in small leasing units. In October, the company extended all the leases for another year. The lessees were actively developing the mine from the Alberta tunnel to the surface. The Cossack tunnel reached 4,500 in length and at last entered the ore zone. The Mine Inspector speculated that, should the ore occur as expected between the new tunnel and the Alberta tunnel 900 feet vertically above it, the new discoveries would add 20 years to the life of the mine.

The Empire was the largest copper producer in the state in 1916. Both oxidized and sulfide ore were shipped to Garfield, Utah, at an average rate of nearly 6,000 tons a month. (The IMIR put production at nearly 8,000 tons per month.) The high metal prices which prevailed during World War I allowed the company to pay out dividends in 1916 that were variously reported as $170,000 and $250,000 (the lower number is probably closer to correct). The Cossack tunnel was 6,000 feet long and the ore in this tunnel was "identical in character with the sulphide mineral shipped from the upper works" (IMIR, p. 29). The upper levels still contained large reserves of ore, even though 1916 had been the most productive year in the mine's history. Over 200 men were employed throughout the year.

In 1917, the Empire was again the largest copper producer in the state, producing more than half of the state's copper output. The mine shipped more than 5,000 tons of ore a month. According to the IMIR, the Empire shipped 70,000 tons of crude ore during 1917 and could have easily produced 100,000 tons, if smelter embargoes and railroad car shortages had not interfered. However, ore shipments were curtailed because the Salt Lake smelters were unable to handle the quantities of ore being shipped from various districts. Development work on the Alberta level discovered the richest and largest sulfide orebody found on the property up to that time. Extensive plans were made for continued development and for installing new equipment. Work was underway to connect the Cossack tunnel to the Alberta tunnel with a 900-foot, three-compartment raise. A vertical four-compartment raise was being driven 500 feet from the Alberta level to the oxidized zone near the surface. A 3-mile-long aerial tramway, planned for a maximum capacity of 125 tons per hour, was under construction. When completed, it was expected to reduce haulage costs 80 percent over the Shay railway. Dividends declared in 1917 amounted to $200,000.

Although the Empire was still the largest copper producer in Idaho in 1918, ore shipments were once again reduced at the request of the smelter. Labor and railroad car shortages caused by the war effort were blamed for the reduced production. Even so, the company shipped over 4,000 tons a month. The aerial tramway, with capacity of 100 tons an hour, began successful operations and the vertical raise above the Alberta tunnel was completed. The raise to connect the Cossack tunnel with the Alberta workings was started, but the work was suspended because the company could not find enough "competent" miners to do the work. A dividend of $50,000 was paid in July.

Ore shipments were greatly curtailed in 1919. The mine shipped only about 1,000 tons of copper ore a month. Considerable time was spent during the year on exploration and development work, as the ore reserves were nearly exhausted. The company sank a 200-foot vertical shaft on the main ore zone from the Copper Bullion tunnel. By the end of the year, the company was drifting on two levels below the Copper Bullion tunnel to locate new reserves. Galena ore was discovered in the Cossack tunnel.

The Empire was again the largest producer in the district in 1920. As usual, most of the ore was mined by lessees, but the company did considerable development work, particularly on the ninth and tenth levels. These levels were reached through the Empire Shaft, an interior shaft that extended downward 340 feet from the Alberta tunnel.

The mine was "practically" idle for most of 1921. In October, the Empire Copper Company was reorganized as Idaho Metals Co. The new company granted many new leases and active operations were resumed. The mine started shipping ore in November and was running at full capacity by the end of the year. In spite of the low activity, the mine was the largest producer in the district. The ore, which was shipped to Garfield, was a mixture of oxide and sulfide minerals.

Worked primarily by lessees, the Empire maintained capacity production of about 1,300 tons (or 40 carloads) per month throughout 1922. The ore consisted of chalcopyrite, bornite, cuprite, and chrysocolla. The company did 3,859 feet of development work and substantially increased ore reserves. The mine had 45 tunnels and 3 shafts. Total development on the property was over 100,000 feet of tunnels, drifts, and crosscuts, and the longest tunnel (the Cossack, on the 1600 level) was 5,500 feet long. The principal tunnels were the Alberta (on the 700 level) and the Cossack. The Cossack tunnel was 900 feet below the Alberta tunnel. The mine was being worked principally through the Empire shaft. Workings also included a glory hole approximately 150 feet in diameter.

In 1923, Idaho Metals Co. continued to operate the Empire through lessees. Ore was shipped to the Garfield smelter at the rate of more than 1,000 tons a month. During the year, 2,400 feet of development work was done and the company had a large quantity of low-grade ore on hand.

Lessees continued active production through the first half of 1924. On July 1, a fire destroyed the mine's ore bins and the headhouse to the aerial tramway. The tramway was also badly damaged by buckets that were set loose and ran out of control during the fire. The tram was repaired, and a new headhouse and ore bins were constructed. Mining operations were resumed on September 25. The company also installed a 150-ton flotation concentrator, which was housed in the old smelter building. The mill was designed to treat the huge volume of low-grade ore that had been located in the mine or that had accumulated in stope fillings over the previous 20 years. It was estimated that it would take 5 years to process the low-grade ore already at hand. Anaconda Copper Mining Co. furnished all the equipment for the mill and was in charge of its installation.

Despite the reduced output due to the fire, the Empire was still the largest producer of copper in Idaho in 1924. Both oxide and sulfide ore were shipped to Utah for smelting. Toward the end of the year, the company also shipped concentrate from the new mill. The company listed the major tunnels as follows: Davis, 1,000 feet; 300 level, 1,800 feet; Alberta, 4,200 feet; Copper Bullion, 2,800 feet; and Cossack, 6,000 feet. The mine had two vertical shafts. One was 560 feet deep, had four compartments, and reached five intermediate levels. The Empire shaft was 320 feet deep, had two compartments, and reached two intermediate levels. Equipment included a 10x10[7] Vulcan air hoist, a 1,500-cubic-foot Laidlaw-Gordon-Dunn compressor, and a 16,300-foot aerial tramway that connected the mine to the railroad and had a capacity of 50 tons per hour. Haulage inside the mine was by mule.

Active production was maintained throughout 1925, and the Empire was again the largest copper producer in Idaho. Over 100 lessees were active at the mine. The mill capacity was increased to 250 tpd, and the company shipped both crude ore and copper concentrate to International, Utah. Much development work was done, but ore reserves were not large. As a result, the mill ran only part time.

The mine was active for only six months in 1926 but was operated all year in 1927. The IMIR noted that the leasing system did not permit sufficient development work, and the company tried to locate financing to explore the Cossack tunnel during 1927. The Empire was the largest copper producer in the state in 1927.

The mine was idle during the first part of 1928, and the property was in the hands of a receiver. In June, the property was purchased from the receivers by Mackay Metals. The new company reconditioned both the mine and mill, started an active development program, and resumed mining with a large number of lessees. By year's end, the company and the lessees were producing enough ore to operate the mill around the clock. The mill treated more than 9,000 tons of copper ore during the year, and the resulting concentrates were shipped to International, Utah, for smelting. The company also shipped more than 1,000 tons of first-class oxidized copper ore. The mine workings totaled over 20 miles of drifts and tunnels, and 1,500 feet of development work was done in 1928.

Besides being the largest copper producer in Idaho in 1929, the Empire ranked third in the state in gold production and was a large producer of silver. The mine and

---

[7]The diameter and stroke, in inches, of the piston that powered the hoist.

29

mill were operated throughout the year. The company did 1,500 feet of development, mostly on the 1,000-foot level and in the Cossack tunnel. A Nordberg 1,250-cubic-foot steam-driven compressor was installed at the Cossack tunnel. Four miles of air line and "a large amount" of new equipment were added to the mine. The 1,000-foot level was opened for production, with the ore consisting chiefly of chalcopyrite, cuprite, and chrysocolla. (Figure 15 shows mine workings at this time.) Nearly 15,800 tons of first-class copper ore was shipped to Utah for smelting. The mill treated 51,000 tons of sulfide ore, producing 4,273 tons of copper concentrates. The copper output for 1929 was more than five times that of 1928.

The Empire Mine and mill were operated continuously during the first part of 1930, but the company suspended all its operations on August 1 because of low copper prices. (Copper reached an all-time low of 0.063 cents per pound in 1932.) Lessees continued to work the mine, but many of them were stockpiling the ore until prices improved. During the year, the mill treated 24,135 tons of copper ore, producing 2,379 tons of concentrates which contained 600 ounces of gold, 14,683 ounces of silver, and 933,529 pounds of copper. In addition, the mine produced 2,032 tons of first-class smelting ore which contained about 4 percent copper. The mine yielded 1,700,000 pounds less than in 1929, but was still the leading producer of copper in Idaho and was ranked sixth in gold production. The company did 1,300 feet of development work. This included advancing the face of the Cossack tunnel to a point 6,000 feet from the portal. Plans, which were never carried out, called for raising a 600-foot raise from the Cossack tunnel to connect with the Alberta tunnel through the Empire shaft.

The property once more passed into receivership on March 28, 1931, due to the Great Depression. Ownership of the property reverted to the county because of unpaid taxes. At the year's end, Mackay Metals had completed arrangements to repurchase the mine and to refinance its debts "as soon as conditions warrant." Five lessees worked the mine during the year, producing a "substantial tonnage" of high-grade ore, which was stored on the property (IMIR, p. 127).

Lessees operated the mine during 1932 and stored the ore while waiting for better metal prices. The 1933 IMIR, citing press reports, stated that the Empire had been sold to M.G. Thomle of Los Angeles, California, who was planning to reopen the mine as soon as metal prices permitted. The rumored sale did not materialize and Mackay Metals continued trying to refinance the mine.

Several small leasing operations worked the upper levels of the mine during 1935. Lessees shipped several carloads of ore from the Empire in 1937, but the activity stopped after word of the pending sale of the property was circulated. G.M. Tomle was rumored to hold an option on the property from the Custer County Commissioners. No changes occurred in the status of the property during 1938. Later reports indicated that Tomle lacked the funds to put the mine into production.

Mackay Exploration acquired the Empire Mine in May 1939. By December, 20 groups of lessees (totaling 45 men) were at work. The company rehabilitated the

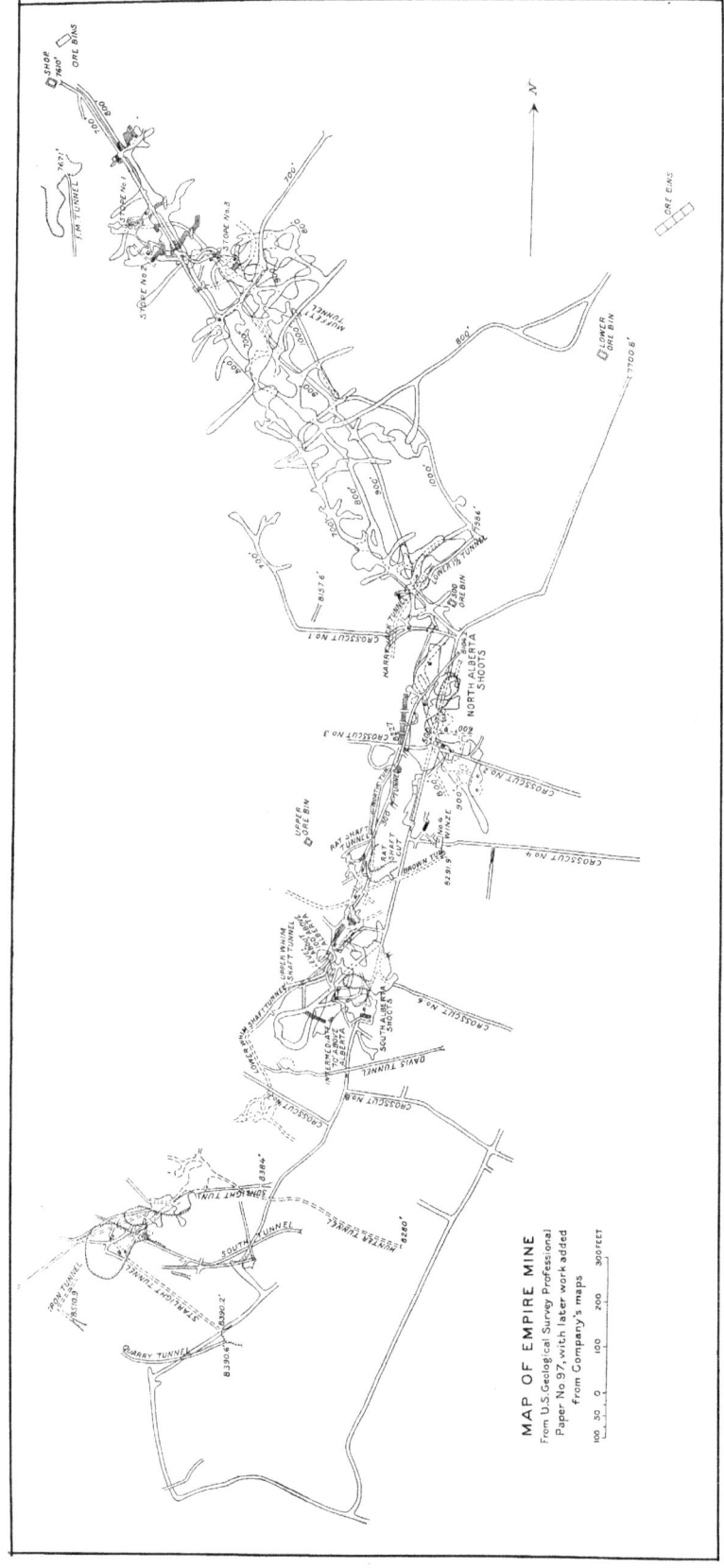

Figure 15. Map of workings of the Empire Mine in 1929 (Figure 2 from The Alder Creek Mining District *in* Ross, Clyde P., 1930).

31

mine and resumed development. The 1,000-foot level was reopened and 480 feet of development was completed by the end of the year. The Cossack tunnel was also reopened and development work started on that level. During 1939, 996 tons of copper ore was shipped to the Salt Lake smelters. About 4,500 tons of ore was shipped in 1940.

The output from the Empire declined to 3,169 tons in 1941. All the production was being done by lessees, who sold the high-grade ores to the Utah smelters but left the mill-grade ore in the mine. The company stated it was working to develop enough low-grade ore to justify starting up milling operations.

The Empire shipped 1,274 tons in 1942. The company also did some development work on tungsten ore and produced 55 tons from a winze below the 1,000 level (Cook, 1956). The mine was closed in August, but in December the U.S. Bureau of Mines started diamond drilling in the mine.

Although the mine was idle for most of 1943, 49 tons of copper ore was shipped. The Bureau operated two diamond drills at the mine during the year with a "satisfactory increase in ore reserves." The U.S. Geological Survey also did some exploration.

In April 1944, Mackay Exploration Co. began reconditioning the mill and power plant, using funds from a $45,000 Reconstruction Finance Corporation (RFC) loan. The company rebuilt the flotation mill, installed machinery at the power plant, and constructed a 2½-mile transmission line from the power plant to the mill. Fifteen men were employed to do this work. Mining, with a work force of 50, began in July. During the second half of the year, 2,330 tons of ore was shipped to the smelter at Garfield and 500 tons was treated in the company mill.

Mackay Exploration operated the mine and 100-ton mill throughout 1945. The mine shipped 6,478 tons of ore (which contained 456 ounces of gold, 7,620 ounces of silver, and 367,893 pounds of copper) to a smelter in Utah. The company also treated 1,850 tons of similar ore by flotation. Some stoping on the orebodies was done during the year.

Custer Copper Corporation, whose president was the former mine manager for Mackay Exploration Co., leased the Empire in 1946. The company made plans to connect the 1,000-foot level to the Cossack tunnel (1,600 level) through a 700-foot inclined raise, eliminating hoisting and outside haulage. (Figure 16 shows the railroad loading station at the lower tramway terminal.) In conjunction with this, the Cossack tunnel was rehabilitated in 1947 and the tracks repaired. The raise was completed in 1948. The mine produced 877 tons of copper ore in 1946, 2,370 tons in 1947, and 431 tons in 1948. Also in 1948, Custer Copper moved 250,000 yards of earth in a surface stripping operation that exposed an ore zone 35 to 40 feet wide containing 6 percent copper.

The Empire shipped 69 tons of ore in 1949. In 1950, work at the mine was confined to surface stripping, opening and sampling an iron ore deposit, and researching sink/float separation of iron and fluorspar ores. The mine was idle except

Figure 16. Railroad loading station at the lower tramway terminal at the Empire Mine (c. 1947) (page 150 from McDowell, George A., 1948, 49th Annual Report of the Mining Industry of the State of Idaho for the Year 1947).

for sampling and maintenance from 1951 to 1953. Two operators shipped ore from the mine in 1954.

In 1955, block lessees worked the Empire Mine and shipped crude ore to a smelter. Late in the year, Idaho Alta Metals Corp., a New York concern, acquired a lease and option on the property. Idaho Alta began a development program with the goal of expanding production from the mine.

Idaho Alta started shipping ore late in October 1956 to the Combined Metals Reduction Co. mill in Bauer, Utah. Work at the mine was delayed by a fire, which destroyed the compressor house and machinery. The company completed 1,700 feet of drifting, 110 feet of raises, 200 feet of diamond drilling, and 18,921 cubic feet of stoping during the year. The main effort involved driving a tunnel 160 feet below the 1,000 level. The tunnel was 1,600 feet long in September 1956, and a raise off this tunnel broke into one of the old Empire stopes, which was nearly 30 feet in diameter (Leland, 1957). About 15 men were employed at the property.

Idaho Alta shipped a substantial quantity of copper ore from the Empire to the Combined Metals Reduction Co. mill in 1957. Output in 1958 was substantially less than in the previous year. The company's lease was terminated in September 1958.

Mackay Exploration Co. produced 593 tons of copper ore from the Empire in 1959, but reported doing only development work. The mine was sold to R.V. Lloyd & Co. on May 10, 1960. The Empire was the only producing mine in the district for the year.

According to the USBM, "sizeable" quantities of copper ore were produced each year from 1961 to 1964. The company built a new flotation mill at the portal of the 1,100 tunnel in 1961 and constructed a new compressor building during 1962. Twelve men were employed in the mining and milling operations during 1963.

U.S. Geological Survey records show that both Idaho Alta and R.V. Lloyd applied for government loans under the Defense Minerals Exploration Administration (Idaho Alta) and Office of Mineral Exploration (R.V. Lloyd) programs. Neither company's application was successful.

Lost River Mines, Inc., purchased the Empire Mine in February 1965. While the company started development work to locate new reserves, ore from the old stopes was mined and milled in the 175-tpd concentrator originally constructed in 1961. A diesel loader and 8-ton ore cars were placed in use on the main 1100 haulage level.

In 1971, the Empire, operated by Honolulu Copper Co., Inc., was active and the property was leased to Ivie Mining Co. Two men were employed on the property. The company employed 10 men in 1972 and 6 men in 1974-1973. Myko, Inc., acquired the Empire mill in 1973-1974, although mining efforts were concentrated on the Horseshoe and Phi Kappa mines. The Empire was active from 1974 through 1977, producing some ore during most of those years. Honolulu Copper employed a work force of 6 men at the mine. Some of the work during this period included operation of an open-pit near the southern end of the property (Figure 17) and construction of a pilot leach plant to test recovery of the oxidized copper ores (Figures 18 and 19).

The mine was inactive during 1977 and 1978. Exxon Corporation drilled the property during that time but decided against further exploration.

In 1980 and 1981, Myko shipped ore from the Phi Kappa Mine to the Empire mill. USBM records show production from the Empire in 1982. However, the distribution of metals (dominantly silver and lead) suggests that this may actually have been Phi Kappa ore that was processed at the Empire mill. (See Figures 20, 21, and 22 for recent views of the mine.) In 1991, Honolulu Copper looked at the area as a potential site for a sulfuric acid copper-leaching operation.

Total recorded production for the Empire Mine between 1902 and 1975 is 921,077 tons of ore. From this were obtained 41,431 ounces of gold, 1,294,531 ounces of silver, 61,689,291 pounds of copper, 24,110 pounds of lead, and 908,078 pounds of zinc. In addition, a small quantity of tungsten ore was shipped in 1942.

BLUE BIRD (EASLIE) MINE

The Blue Bird Mine (Easlie Group) is located in Rio Grande Canyon just north of the Empire Mine (Figures 2, 3, and 22). When Umpleby (1917) visited the district in 1912, the Easlie group was developed by a shaft and several short tunnels, all of which were inaccessible. The claims covered a large limestone mass surrounded by granite porphyry. The tunnels followed the east contact between the limestone and the granite, while the shaft and several prospect pits were on the west contact. A small carload of ore, containing 30 percent lead and between 8 and 9 ounces of silver per ton, was produced in 1909 (Umpleby, 1917).

In 1924, activities on the Darlington and Fowler lease at the "Blue Jay" (the name by which the IMIR incorrectly referred to the property) added considerably to the activity in the district. The lessees did about 50 feet of development work, and some oxidized lead ore was shipped from the property during the year. The mine had a 100-foot vertical shaft and about 200 feet of drifts.

Lessees made important ore discoveries during 1925, and several lots of ore were shipped. In 1928, a lessee shipped a few hundred tons of smelting-grade sulfide lead ore and sent one carload of lead-zinc ore to a custom flotation mill at Midvale, Utah. (The reports of activity at the Blue Jay, "owned by the White Knob Mining Co.," in the 1927 IMIR probably refer to the Blue Bell, one of the claims that was owned by White Knob.)

The mine was inactive from 1929 to 1935. In 1936, production from the Blue Bird was credited with increasing the output of lead-silver ore from the district. The 1937 IMIR mentioned production from the "Blue Bell" (another incorrect reference by the Mine Inspector) by lessees Crocker and Judd. The Blue Bird produced ore in 1938

Figure 18. Leach pad and staging area at the open-pit mine (1994) (Idaho Geological Survey photograph by Falma J. Moye).

Figure 19. Cement leach pad and copper plating system at the open-pit copper mine (1994) (Idaho Geological Survey photograph by Falma J. Moye).

Figure 20. Looking south at the main level portals of the Empire Mine (c. 1990) (Figure 13 from McHugh and others. 1991).

Figure 21. The Empire Mine (c. 1985). The Lost River Range forms the distant skyline (page 181 from Link, P.K., and W.R. Hackett, editors, Guidebook to the Geology of Central and Southern Idaho: Idaho Geological Survey Bulletin 27).

40

Figure 22. Overview of the Empire Mine area, with the White Knob Mountains in the background. An ore bin and tram station are in the center of the photograph, and the open-pit workings near the Blue Bird Mine are the reddish area to the upper right (Idaho Geological Survey photograph by Falma J. Moye).

41

and 1939. During 1938, Mackay Metals Consolidated was organized to consolidate the Blue Bird and several other properties adjoining the Empire Mine. According to the 1938 IMIR, the company was waiting for approval from the Securities and Exchange Commission before starting operations. Later IMIRs make no mention of this company.

According to McHugh and others (1991), the main shaft on the Blue Bird was at least 100 feet deep. Other development on the property consisted of several pits and short adits. Between 1918 and 1939, the property produced 1,530 tons of ore. From this was obtained 23 ounces of gold, 16,426 ounces of silver, 12,595 pounds of copper, 509,165 pounds of lead, and 3,111 pounds of zinc.

CHAMPION MINE

The Champion Mine was variously reported to have been located in 1895 (Nelson and Ross, 1968) and 1901 (Umpleby, 1917). In 1912, the Champion group consisted of nine claims located south-southeast of the Empire Mine at an elevation of about 8,000 feet (Figures 2 and 3). According to Umpleby, the workings consisted of three main tunnels in the central part of the property and several small tunnels located toward the north end. He noted that the Champion was the most important deposit in the district to be discovered outside the main body of granite porphyry. The workings on the property were almost entirely within limestone near the contact with the granite.

The first recorded production from the Champion was in 1908. Little mention was made of the property in the following years, but Champion Mining Co. (incorporated on November 9, 1905) consistently reported doing assessment work on the claim group. (See Table 5 for development work done on the mine.)

The company's report for 1914 mentioned several tunnels, the longest of which was 470 feet long. Most of the work done during the year was clearing and retimbering caved workings. Total development at the mine was about 1,550 feet of workings. A lessee shipped 8 tons of sorted ore during the year; the company stated the average value of this ore was $15.49 per ton, and that 48 ounces of silver, 3,200 pounds of lead, and 800 pounds of zinc were recovered. Company reports from 1914 to 1917 stated most of the development work was being done on the O.S.L. claim, partly with contract workers, and that assessment work was being done to maintain the claim group (Figure 23).

In October 1917, the entire property was placed under lease. Total development at that time was approximately 1,800 feet of workings. It is not known when the lease was terminated; company reports do not mention lessees after 1920.

The 1922 company report described the mine workings as consisting of nine tunnels, five winzes, three cross-cuts, and five drifts. The No. 1 tunnel was 100 feet long; the No. 2, 470 feet; the No. 3, 275 feet; the No. 4, 85 ft; and the No. 5, 40

42

Table 5. Development work and number of employees at the Champion Mine.

Year	No. of Men employed	Tunnels (feet)	Sinking (feet)	Cross-cutting (feet)	Operator
1914	2	10[1]	---	---	Champion Mining Co. and lessee
1915	2	40[2]	---	40[3]	Champion Mining Co.
1916	[4]	100[2]	8[5]	90[3]	Champion Mining Co. and contract workers
1917	2	75[2]	25[5]	50[3]	Champion Mining Co. and contract workers
1919	[6]	68[2]	---	68[3]	Lessees
1920	[6]	65[2]	---	65[3]	Lessees
1921	[7]	40[2]	---	40[3]	Champion Mining Co.
1922	1	50[8]	---	---	Champion Mining Co.
1923	1	25[8]	---	---	Champion Mining Co.
1924	1	35[9,10]	---	---	Champion Mining Co.
1925	1	35[11]	---	---	Champion Mining Co.
1926	1	50[10]	---	---	Champion Mining Co.

[1]Work for the year consisted mostly of clearing caved workings and retimbering these areas.
[2]Total development work for the year.
[3]Combined figure for cross-cutting and drifting.
[4]Company stated that work was done mainly on contract. Number of workers not reported.
[5]Combined figure for sinking and raising.
[6]Work done by lessees. Number of workers not reported.
[7]Number of workers not reported.
[8]Total development work for the year. Included drifting, cross-cutting, and raising.
[9]Total development work for the year. Included drifting and cross-cutting.
[10]Work for the year also included cleaning out old tunnels, retimbering, and some surface work.
[11]Work also included some surface and repair work.

feet. The "No. 5 tunnel" was located at the cabin and was used as a root cellar. Total mine workings were about 2,000 feet, consisting of 150 feet of winzes, 75 feet of raises, and 1,775 feet of tunnels, cross-cuts, and drifts.

Minor work was done on the property between 1922 and 1926. Total development in 1926 was 2,100 feet, which included 150 feet of winzes, 75 feet of inclined raises, and 1,900 feet of tunnels, cross-cuts, and drifts. The company noted that a new tunnel had been started at the south end of the property, headed toward the main mineralized areas of the mine. The Champion Mining Company forfeited its charter in 1927.

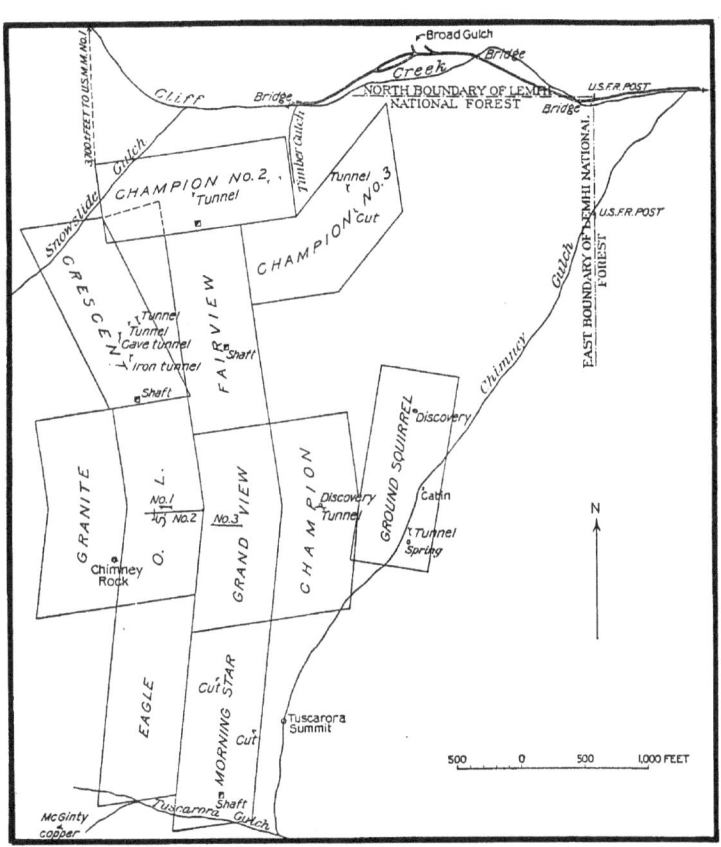

Figure 23. Claim map of the Champion Group (Figure 13 from Umpleby, 1917).

The next mention of the Champion mine was in 1943. The mine was active for most of the next decade. In 1943, 322 tons of lead ore was shipped to a smelter and 200 tons in 1944. The mine also shipped ore in 1947, 1948, and 1949. It was the principal producer in the district in 1951 and was active in 1952. In 1954, lessees produced 282 tons of ore, which yielded 53 tons of lead concentrate and 12 tons of zinc concentrate. A "substantial tonnage" of lead ore was produced in 1955.

U.S. Geological Survey records show that Joseph Ausich applied for government assistance under both the Defense Minerals Exploration Administration (DMEA) and the Office of Mineral Exploration (OME) programs. A DMEA contract for $31,650 was awarded on June 30, 1952; government participation in the project was 50 percent. The total amount of money was increased to $39,510 in 1954. The project was active from 1952 to 1954, but apparently no significant orebodies were discovered.

The mine produced ore in 1962. Assessment work was done on the property by Joe Ausich and/or Ausich Mines during the early 1970s. The mine workings total about 3,032 feet of drifts, crosscuts, raises, and winzes on three levels (McHugh and others, 1991; Figure 24). Production for the mine, between 1937 and 1970, was 2,708 tons of ore averaging 0.0069 ounce of gold per ton, 2.76 ounces of silver per ton, 0.3 percent copper, 12.8 percent lead, and 2.8 percent zinc. Between 1908 and 1964 the mine produced 2,281 tons of ore, which yielded 14 ounces of gold, 7,236 ounces of silver, 20,003 pounds of copper, 579,668 pounds of lead, and 32,719 pounds of zinc.

DOUGHBOY MINE

The Doughboy Mine is located in Cougar Gulch about two miles northwest of the Empire (Figure 2). By 1919, the mine was developed by a 100-foot-deep inclined shaft driven along the contact between the limestone and an intrusive porphyry. The mine shipped 3 carloads of ore averaging 30 to 40 percent lead and nearly an ounce of silver per unit of lead.

The Doughboy Mining and Leasing Co. (T.M. Douglass, Jr., Assistant Secretary) was incorporated on May 7, 1920. The company forfeited its charter in December 1921.

The 1920 company report to the Mine Inspector listed development on the property as a 243-foot shaft and 400 feet of drifting. A large shipment of ore was made in 1920, and smaller shipments were made in 1921 and in 1923 through 1925. Small amounts of ore were produced in 1943 and 1944. The mine was reopened in the 1960s, and the property produced 6 tons of lead in 1962.

According to McHugh and others (1991), the mine was developed by one 370-foot shaft (now caved), 1,500 feet of drifting, one short adit, and one pit. Between

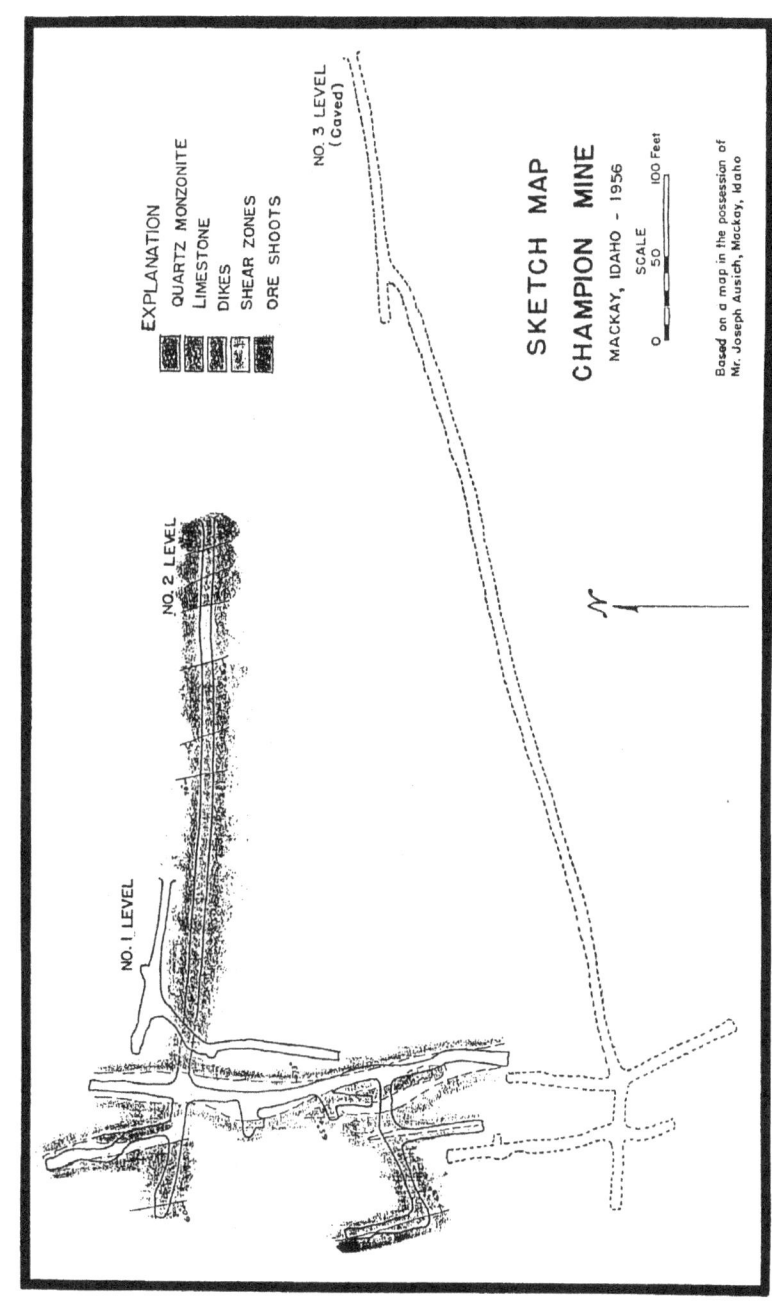

Figure 24. Sketch map of the workings of the Champion Mine (Figure 7 from Leland, 1957).

46

1919 and 1968, the mine produced 1,070 tons of ore. This yielded 8 ounces of gold, 31,703 ounces of silver, 1,972 pounds of copper, 637,120 pounds of lead, and 3,900 pounds of zinc.

HORSESHOE MINE

The Horse Shoe Copper Company reports to the Idaho Mine Inspector indicate that the company acquired the Horseshoe Group in 1903. (Table 6 lists the companies operating at the mine). When Umpleby (1917) visited the Alder Creek district in 1912, the Horseshoe group consisted of about thirteen unpatented claims located north and east of the Empire (Figures 2 and 3). The property was developed by a few open cuts and short tunnels, but it had produced no ore.

The first recorded production from the Horseshoe was in 1916, when the mine made a few shipments of lead ore. The IMIR (p. 28) noted that lessees had discovered "a splendid shoot of desirable lead carbonate and galena ore," from which two carloads of ore yielded the miners a "good margin of profit."

In 1917, lessees at the Horseshoe shipped nearly 2,000 tons of ore from shallow, near-surface workings. The ore ran over 29 percent lead and about 10 ounces of silver per ton.

In 1918 and 1919, the property was worked by U.S. Smelting and Refining Co. of Salt Lake City. The mine shipped ore in both years. Output for 1919 totaled about 1,000 tons of ore which ran about 30 percent lead and 30 ounces of silver per ton. The 1919 IMIR described the ore (p. 82) as consisting of: ". . . a lenzy contact deposit of crystallized lead and sand carbonate ore between limestone and granite porphyry. These lenzes have varied up to 20 feet thick, but have proven decidedly irregular so far in both length and depth but of persistent recurrence to a point nearly 200 feet below the outcrop and have made a handsome total yield of desirable mineral from previous leasing operations."

A large quantity of lead ore was shipped from the Horseshoe during 1920. Lessees discovered a new ore zone during 1921, and a number of shipments of lead ore were made, although the output was small compared to previous years. "Considerable" oxidized lead ore was shipped to smelters in Utah during 1922 (USGS). Total workings for the mine were said to be "Many thousand feet". The principal workings consisted of three tunnels, two short shafts, and numerous crosscuts, drifts, and raises. A large amount of development work was done during the year.

In 1923, the company placed the mine under a 3-year lease to Wayne Darlington. Standard lease terms included a 10 percent royalty. Darlington also had the option to purchase the mine at 7 cents per share (about $70,000 total). In addition to shipping ore, Darlington opened a new orebody of chalcopyrite and pyrrhotite.

47

Table 6. Companies operating at the Horseshoe Mine.

Company Name	Officer	Date Incorporated	Charter Forfeited	Year(s) at Mine
Horse Shoe Copper Co., Ltd.	George L. Morgan, President/Manager	Dec. 10, 1903	Dec. 1, 1930	1903-1930
Kay Development Co., Inc.	Wayne Darlington, President/Manager	March 17, 1924	1	1924-1928
U.S. Silver Corp.	1	1	taken over by Myko, Inc.	1971-1972
Myko, Inc.	Ivan Taylor, Vice President	March 7, 1973	not reported as active in 1981	1974-1976

[1]Information not available in IGS's files.

In 1924, Darlington's Kay Development Co., Inc., assumed control of the property. During the year, the company constructed a 4-mile electric transmission line from Mackay to the mine, installed new mine equipment, and conducted an active development program which included sinking the main shaft an additional 100 feet. (See Table 7 for development work done at the mine.) Several lots of ore were shipped. According to the company, most of the old workings in the mine were caved.

Kay Development conducted active operations throughout 1925. The main shaft was sunk to a deeper level and considerable lateral exploration was done on the mine's upper levels. The workings consisted of two tunnels and a two-compartment vertical shaft with four intermediate levels. The No. 1 tunnel was 450 feet long, and the No. 2 tunnel was 600 feet long. The mine was equipped with an electric hoist and an electrically driven two-drill air compressor.

Early in 1926, the company opened up considerable new ore. Production for the year was nearly 1,000 tons of silver-lead ore. In addition, some lead-zinc ore was shipped to a custom flotation plant for testing.

Active development work during the early part of 1927 located a large body of complex lead-zinc ore. The company's 1928 report to the Idaho Mine Inspector stated that the mine's surface plant and equipment had been destroyed by a snow slide. Kay Development shipped one carload of copper-lead-zinc ore to a custom flotation plant during 1928. The IMIR entry for the Horse Shoe Copper Co. said the property had been leased to Ray Strunk of Mackay.

Lessees did a small amount of work on the mine during 1930. Total workings on the property were over 2,000 feet (Figure 25). Horse Shoe Copper Co., Ltd., forfeited its charter on December 1, 1930.

Table 7. Development work, number of men employed, and operating companies at the Horseshoe Mine.

Year	No. of Men employed	Tunnels (feet)	Sinking (feet)	Cross-cutting (feet)	Drifting (feet)	Raising (feet)	Operator
1924	12	---	100	400	---	---	Kay Development Co.
1925	7	---	90	100	200	---	Kay Development Co.
1926	6	---	---	150	350	---	Kay Development Co.
1928	1[1]	---	---	---	---	100	Kay Development Co.
1931	2	150	---	---	---	---	Horseshoe Copper Co., Ltd.

[1]Number of men employed was not reported.

Lessees Whitney and Anderson shipped ore from the Horseshoe in 1937. The mine was operated from 1937 to 1951, and ore was shipped every year except 1942. The Horseshoe shipped 78 tons of zinc-lead ore in 1943, 252 tons of zinc-lead ore in 1944, 175 tons of silver-lead ore in 1945, and 244 tons of lead ore in 1946. In 1950, the mine shipped 75 tons of lead smelting ore. A small quantity of lead ore was produced from the mine by Ira C. Lambert and V.A. Anderson in 1954.

U.S. Silver Corporation operated the mine during 1971 and 1972. The company had eight men working during 1972. Myko, Inc., worked the property from 1974 to 1976 with a force of three or four men.

According to Nelson and Ross (1968), the mine was developed on six levels, of which the deepest was the 350 level. In 1929, the lower levels and most of the stopes were filled with water (Ross, 1930). The property was examined in 1955 for a DMEA loan, but no contract was awarded. The two adits on the property were about 875 and 1,225 feet long. Workings on the four lower levels, reached through a 350-foot shaft, totaled about 1,500 feet (McHugh and others, 1991).

Production from 1916 through 1978 was 13,916 tons of ore which yielded 108 ounces of gold, 149,461 ounces of silver, 37,350 pounds of copper, 4,186,963 pounds of lead, and 446,014 pounds of zinc. In 1979 the mine shipped 2,900 tons of ore, which averaged 3.8 percent copper and 12 percent zinc, to the Empire mill. Work in the late 1980s exposed a body of sulfide ore containing 12-14 percent zinc, 5-6 percent lead, and 0.7-1.0 ounce per ton of silver (McHugh and others, 1991).

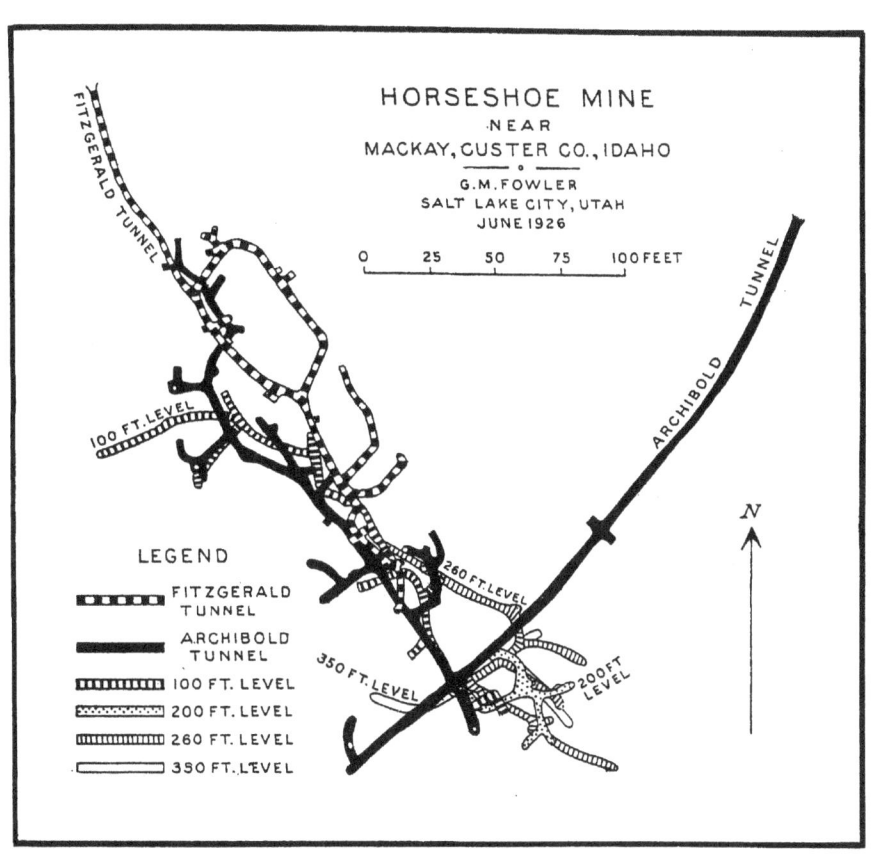

Figure 25. Workings of the Horseshoe Mine (Figure 3 from The Alder Creek Mining District *in* Ross, 1930).

50

WHITE KNOB GROUP

The White Knob Group (not to be confused with the earlier White Knob Mine that became the Empire) was named after the company formed in 1919 to operate the mine. The main claims in the group were the Homestake, Copper Queen, and Blue Bell, which were located in Rio Grande Canyon about half a mile northwest of the main workings of the Empire Mine (Figure 2). The first record of the property was in 1909, when a small shipment of ore was made from the Homestake. The Homestake also produced ore in 1911. Both the Homestake and the Copper Queen were credited with producing "carbonate lead ore" in 1915. In 1916 the two claims shipped about 60 cars of lead ore, and in 1917 production averaged nearly 1,000 tons a month.

In the early part of 1918, the property was bonded by the United States Smelting, Refining & Mining Co., of Salt Lake City, Utah. Several thousand tons of oxidized lead ore was produced during the year.

The White Knob Mining Company (controlled by United States Smelting, Refining & Mining Co.) was organized in 1919. The mine was idle that year but shipped a "large" amount of ore in 1920. Production for 1921 was greatly reduced from that of the previous year. The mine had 5,000 feet of underground workings. The company did 800 feet of development work during the year. (See Table 8 for development work done at the mine.)

The mine, operated mostly by lessees, shipped ore every year from 1922 to 1929. The White Knob shipped nearly 2,000 tons of oxidized lead ore to the smelter at Midvale in 1923. The mine shipped several hundred tons of oxidized lead ore in 1924. Shipments were somewhat smaller in 1925, but lessees located important new ore reserves. Several cars of oxidized lead ore were produced in 1926, but shipments for 1927 to 1929 were one about one carload per year. In 1928, the mine had two tunnels, one shaft, twenty-two raises, six cross-cuts, and seven drifts. The No. 1 tunnel was 1,100 feet long, the No. 2 tunnel was 900 feet long, and the shaft was 250 feet deep. The mine had a 40-horsepower gasoline hoist.

Lessees worked the mine and shipped at least some ore every year from 1937 to 1943. In 1938, the company did 347 feet of development work on the new No. 3 tunnel. The mine had one 1-ton ore car and 300 feet of track, which were used by the lessees. The company described the hoist as a 16-horsepower Western gas engine, but went on to say that it was "Partly dismantled and not in use". Two lessees worked the mine in 1939.

Development work for 1940 included "Tunnels, 175 ft.; Sinking, 125 ft.; Drifting, 185 ft.; Raising, 55 ft." (IMIR, p. 132). The company's report to the Mine Inspector mentioned two new tunnels. The No. 4 tunnel was 100 feet long, and the No. 5 tunnel was 75 feet long. Twenty feet of work was done on the No. 6 tunnel in 1942, and 50 feet were driven on the new No. 7 tunnel in 1943.

Table 8. Development work and number of men working at the White Knob Mine.

Year	No. of Men employed[1]	Tunnels (feet)	Sinking (feet)	Cross-cutting (feet)	Drifting (feet)	Raising (feet)	Operator
1921	7	800[2]	150[3]	650[4]	---	---	White Knob Mining Co.
1938[5]	[6]	347	6	---	---	---	White Knob Mining Co.
1940	[6]	175	125	---	185	55	White Knob Mining Co.
1941	[6]	---	50	---	130	175	White Knob Mining Co.
1942	3	20	---	---	---	---	White Knob Mining Co.
1943	8	50[7]	---	---	---	---	White Knob Mining Co.
1946	3	---	30	145	190	---	White Knob Mining Co.
1947	9	[8]	33	259	310	---	White Knob Mining Co.
1948	18	330	100	---	175	---	White Knob Mining Co.
1949	20	---	150	---	790	---	White Knob Mining Co.
1950	11	---	100	---	591	---	White Knob Mining Co.
1951	4	---	8	---	191	56	White Knob Mining Co.
1952	2	---	42	---	136	---	White Knob Mining Co.
1953	2	---	27	---	25	---	White Knob Mining Co.
1954	2	---	24	---	76	45	White Knob Mining Co.
1955	2	62	14	---	---	30	White Knob Mining Co.
1961	2	[9]	71	---	50	30	White Knob Mining Co.
1962	2	---	15	---	---	---	White Knob Mining Co.
1964	2	---	35	---	15	---	White Knob Mining Co.

[1]"Number of men employed" includes lessees. For most years, development work was done entirely by lessees.
[2]Total development work for the year.
[3]Combined figure for sinking and raising.
[4]Combined figure for cross-cutting and drifting.
[5]Mine was inactive from 1930 to 1937. Number of workers and amount of development not reported for years that mine was active between 1921 and 1938.
[6]Number of lessees working the mine was not reported.
[7]Development work for the year also included an open cut 225 feet long and 15 feet wide, ranging in depth from zero to 20 feet deep.
[8]Development work for the year also included work on the open cut. Dimensions were increased to 275 feet in length, and 50 to 70 feet wide at the top and 25 to 40 feet wide at the bottom.
[9]Development work for the year also included 50 feet of work on an open cut.

In 1942, lessees at the White Knob shipped 2,109 tons of crude zinc ore to the International slag-fuming plant at Tooele, Utah, between January and July. They also shipped 46 tons of silver-lead ore to a smelter.

The mine was idle in 1943 and 1944 but made a small shipment in 1945. The Homestake was the most important producer in the district in 1946. The mine shipped 5,605 tons of ore to a lead smelter where the gold, silver, and lead were recovered. The hot slag was then fumed to recover the zinc. In addition, 171 tons of silver-lead ore was shipped directly to a lead smelter. Development work for the year, done by lessees, consisted of 30 feet of sinking, 145 feet of cross-cutting, and 190 feet of drifting.

In 1947, the mine produced 4,979 tons of lead-silver ore and 1,680 tons of zinc ore. Production for 1948 was 5,082 tons of lead-silver ore and 711 tons of zinc ore. An average of 18 lessees worked the mine. Output for 1949 was 3,171 tons of lead-silver ore and 405 tons of zinc ore. Shipments of ore were also made in 1950, 1951, 1952, and 1954. In 1951 mine equipment included an 8x10[8] M&S steam hoist and a Gardner-Denver portable air compressor and air receiver. Lessees Myron and Curt Fullmer shipped lead ore to the Midvale, Utah, smelter in 1954.

According to the company, lessees worked the mine in 1955, but no ore was shipped. The mine was idle from 1956 to 1959. Lessees operated the mine from 1960 to 1969. Small shipments of ore were made in most years, although only assessment work was done for 1965 and 1966. Reports for the years after 1969 either list the mine as idle or mention that assessment work was done for the year. The 1974-1975 IMIR listed UV Industries as the controlling company for White Knob Mining Co. (replacing U.S. Smelting & Refining).

U.S. Geological Survey records show an application was made to the Defense Minerals Exploration Administration for assistance in exploring the Copper Queen. The application was not successful.

Total production for the mine between 1909 and 1968 was 51,501 tons of ore. From this 409 ounces of gold, 295,300 ounces of silver, 252,274 pounds of copper, 9,455,350 pounds of lead, and 3,436,438 pounds of zinc were obtained.

References

Cook, E.F., 1956, Tungsten deposits of south-central Idaho: Idaho Bureau of Mines and Geology Pamphlet 108, 40 p.

Farwell, F.W., and R.P. Full, 1944, Geology of the Empire copper mine near Mackay, Idaho: U.S. Geological Survey Open-File Report 44-17, 22 p.

---

[8]The diameter and stroke, in inches, of the piston which powered the hoist.

Idaho Geological Survey's mineral property files (includes copies of company reports to the Idaho Inspector of Mines).

Idaho Geological Survey's (IGS) annual reports on Regional Developments in Minerals, Mining, and Energy in Idaho, 1975-1992.

Idaho Inspector of Mines' annual reports (IMIR) on the Mining Industry of Idaho, 1899-1970.

Kemp, J.F., and C.G. Gunther, 1908, The White Knob copper-deposits, Mackay, Idaho: American Institute of Mining Engineers, v. 38, 269-296.

Leland, G.R., 1957, General geology and mineralization of the Mackay stock area: University of Idaho M.S. thesis, 71 p.

McHugh, E.L., H.W. Campbell, M.C. Horn, and T.J. Close, 1991, Mineral resource appraisal of the Challis National Forest, Idaho: U.S. Bureau of Mines Mineral Land Assessment Open-File Report MLA 6-91, 319 p.

Nelson, W.H., and C.P. Ross, 1986, Geology of part of the Alder Creek mining district, Custer County, Idaho: U.S. Geological Survey Bulletin 1252-A, 30 p.

Ross, C.P., 1930, Geology and ore deposits of the Seafoam, Alder Creek, Little Smoky and Willow Creek mining districts, Custer and Camas counties, Idaho: Idaho Bureau of Mines and Geology Pamphlet 33, 26 p.

Umpleby, J.B., 1917, Geology and ore deposits of the Mackay region, Idaho: U.S. Geological Survey Professional Paper 97, 129 p.

U.S. Geological Survey (USGS)/U.S. Bureau of Mines (USBM) Minerals Yearbook chapters for Idaho, 1900-1990.

Wells, M.W., 1983, Gold camps & silver cities: nineteenth century mining in central and southern Idaho: Idaho Bureau of Mines and Geology Bulletin 22, 165 p.

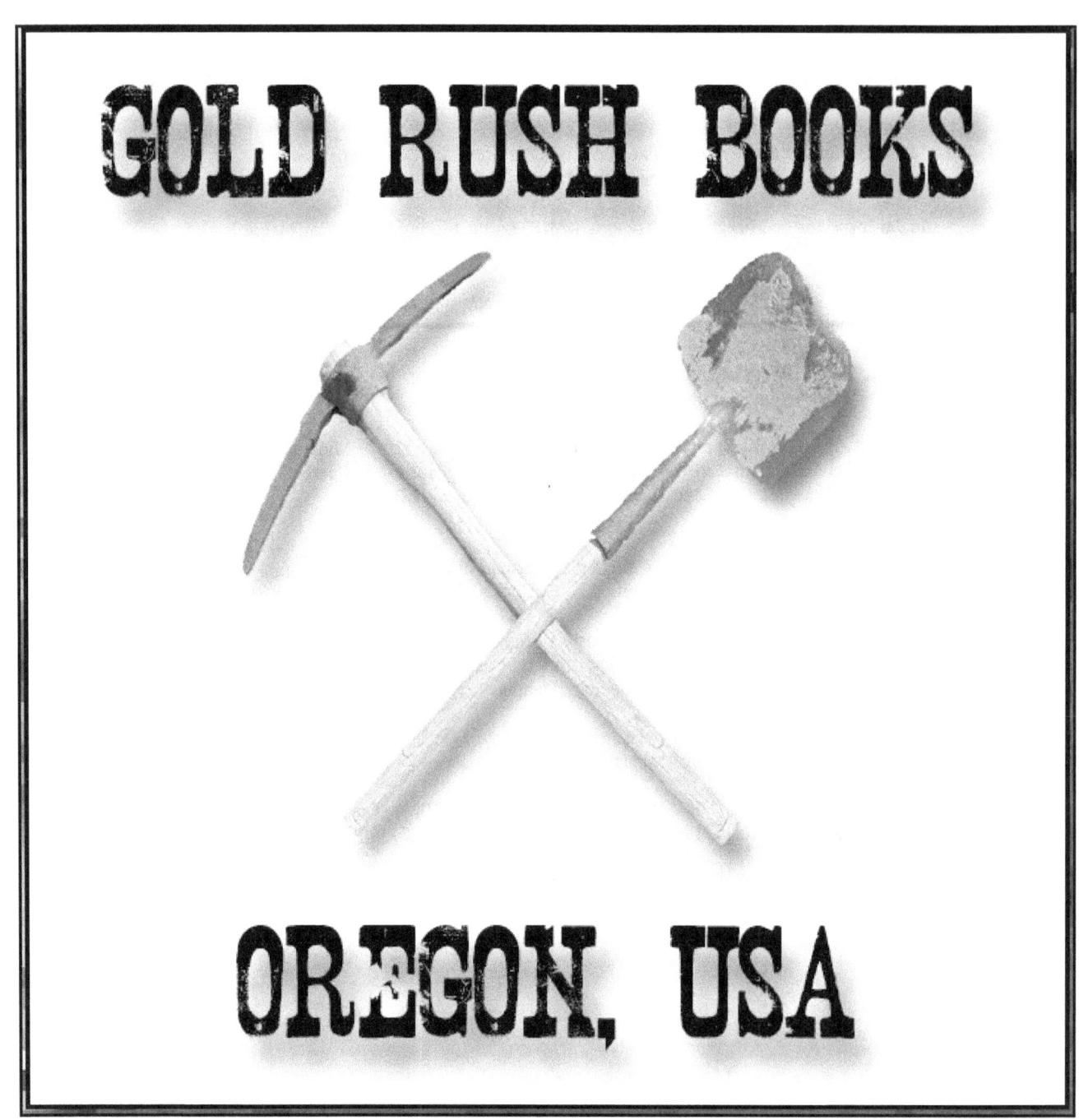

GOLD RUSH BOOKS

OREGON, USA

www.GoldMiningBooks.com

Books On Mining

Visit: www.goldminingbooks.com to order your copies or ask your favorite book seller to offer them.

Mining Books by Kerby Jackson

<u>Gold Dust: Stories From Oregon's Mining Years</u> - Oregon mining historian and prospector, Kerby Jackson, brings you a treasure trove of seventeen stories on Southern Oregon's rich history of gold prospecting, the prospectors and their discoveries, and the breathtaking areas they settled in and made homes. 5" X 8", 98 ppgs. Retail Price: $11.99

<u>The Golden Trail: More Stories From Oregon's Mining Years</u> - In his follow-up to "Gold Dust: Stories of Oregon's Mining Years", this time around, Jackson brings us twelve tales from Oregon's Gold Rush, including the story about the first gold strike on Canyon Creek in Grant County, about the old timers who found gold by the pail full at the Victor Mine near Galice, how Iradel Bray discovered a rich ledge of gold on the Coquille River during the height of the Rogue River War, a tale of two elderly miners on the hunt for a lost mine in the Cascade Mountains, details about the discovery of the famous Armstrong Nugget and others. 5" X 8", 70 ppgs. Retail Price: $10.99

Oregon Mining Books

<u>Geology and Mineral Resources of Josephine County, Oregon</u> - Unavailable since the 1970's, this important publication was originally compiled by the Oregon Department of Geology and Mineral Industries and includes important details on the economic geology and mineral resources of this important mining area in South Western Oregon. Included are notes on the history, geology and development of important mines, as well as insights into the mining of gold, copper, nickel, limestone, chromium and other minerals found in large quantities in Josephine County, Oregon. 8.5" X 11", 54 ppgs. Retail Price: $9.99

<u>Mines and Prospects of the Mount Reuben Mining District</u> - Unavailable since 1947, this important publication was originally compiled by geologist Elton Youngberg of the Oregon Department of Geology and Mineral Industries and includes detailed descriptions, histories and the geology of the Mount Reuben Mining District in Josephine County, Oregon. Included are notes on the history, geology, development and assay statistics, as well as underground maps of all the major mines and prospects in the vicinity of this much neglected mining district. 8.5" X 11", 48 ppgs. Retail Price: $9.99

<u>The Granite Mining District</u> - Notes on the history, geology and development of important mines in the well known Granite Mining District which is located in Grant County, Oregon. Some of the mines discussed include the Ajax, Blue Ribbon, Buffalo, Continental, Cougar-Independence, Magnolia, New York, Standard and the Tillicum. Also included are many rare maps pertaining to the mines in the area. 8.5" X 11", 48 ppgs. Retail Price: $9.99

<u>Ore Deposits of the Takilma and Waldo Mining Districts of Josephine County, Oregon</u> - The Waldo and Takilma mining districts are most notable for the fact that the earliest large scale mining of placer gold and copper in Oregon took place in these two areas. Included are details about some of the earliest large gold mines in the state such as the Llano de Oro, High Gravel, Cameron, Platerica, Deep Gravel and others, as well as copper mines such as the famous Queen of Bronze mine, the Waldo, Lily and Cowboy mines. This volume also includes six maps and 20 original illustrations. 8.5" X 11", 74 ppgs. Retail Price: $9.99

<u>Metal Mines of Douglas, Coos and Curry Counties, Oregon</u> - Oregon mining historian Kerby Jackson introduces us to a classic work on Oregon's mining history in this important re-issue of Bulletin 14C Volume 1, otherwise known as the Douglas, Coos & Curry Counties, Oregon Metal Mines Handbook. Unavailable since 1940, this important publication was originally compiled by the Oregon Department of Geology and Mineral Industries includes detailed descriptions, histories and the geology of over 250 metallic mineral mines and prospects in this rugged area of South West Oregon. 8.5" X 11", 158 ppgs. Retail Price: $19.99

Metal Mines of Jackson County, Oregon - Unavailable since 1943, this important publication was originally compiled by the Oregon Department of Geology and Mineral Industries includes detailed descriptions, histories and the geology of over 450 metallic mineral mines and prospects in Jackson County, Oregon. Included are such famous gold mining areas as Gold Hill, Jacksonville, Sterling and the Upper Applegate. **8.5" X 11", 220 ppgs. Retail Price: $24.99**

Metal Mines of Josephine County, Oregon - Oregon mining historian Kerby Jackson introduces us to a classic work on Oregon's mining history in this important re-issue of Bulletin 14C, otherwise known as the Josephine County, Oregon Metal Mines Handbook. Unavailable since 1952, this important publication was originally compiled by the Oregon Department of Geology and Mineral Industries includes detailed descriptions, histories and the geology of over 500 metallic mineral mines and prospects in Josephine County, Oregon. **8.5" X 11", 250 ppgs. Retail Price: $24.99**

Metal Mines of North East Oregon - Oregon mining historian Kerby Jackson introduces us to a classic work on Oregon's mining history in this important re-issue of Bulletin 14A and 14B, otherwise known as the North East Oregon Metal Mines Handbook. Unavailable since 1941, this important publication was originally compiled by the Oregon Department of Geology and Mineral Industries and includes detailed descriptions, histories and the geology of over 750 metallic mineral mines and prospects in North Eastern Oregon. **8.5" X 11", 310 ppgs. Retail Price: $29.99**

Metal Mines of North West Oregon - Oregon mining historian Kerby Jackson introduces us to a classic work on Oregon's mining history in this important re-issue of Bulletin 14D, otherwise known as the North West Oregon Metal Mines Handbook. Unavailable since 1951, this important publication was originally compiled by the Oregon Department of Geology and Mineral Industries and includes detailed descriptions, histories and the geology of over 250 metallic mineral mines and prospects in North Western Oregon. **8.5" X 11", 182 ppgs. Retail Price: $19.99**

Mines and Prospects of Oregon - Mining historian Kerby Jackson introduces us to a classic mining work by the Oregon Bureau of Mines in this important re-issue of The Handbook of Mines and Prospects of Oregon. Unavailable since 1916, this publication includes important insights into hundreds of gold, silver, copper, coal, limestone and other mines that operated in the State of Oregon around the turn of the 19th Century. Included are not only geological details on early mines throughout Oregon, but also insights into their history, production, locations and in some cases, also included are rare maps of their underground workings. **8.5" X 11", 314 ppgs. Retail Price: $24.99**

Lode Gold of the Klamath Mountains of Northern California and South West Oregon
(See California Mining Books)

Mineral Resources of South West Oregon - Unavailable since 1914, this publication includes important insights into dozens of mines that once operated in South West Oregon, including the famous gold fields of Josephine and Jackson Counties, as well as the Coal Mines of Coos County. Included are not only geological details on early mines throughout South West Oregon, but also insights into their history, production and locations. **8.5" X 11", 154 ppgs. Retail Price: $11.99**

Chromite Mining in The Klamath Mountains of California and Oregon
(See California Mining Books)

Southern Oregon Mineral Wealth - Unavailable since 1904, this rare publication provides a unique snapshot into the mines that were operating in the area at the time. Included are not only geological details on early mines throughout South West Oregon, but also insights into their history, production and locations. Some of the mining areas include Grave Creek, Greenback, Wolf Creek, Jump Off Joe Creek, Granite Hill, Galice, Mount Reuben, Gold Hill, Galls Creek, Kane Creek, Sardine Creek, Birdseye Creek, Evans Creek, Foots Creek, Jacksonville, Ashland, the Applegate River, Waldo, Kerby and the Illinois River, Althouse and Sucker Creek, as well as insights into local copper mining and other topics. **8.5" X 11", 64 ppgs. Retail Price: $8.99**

Geology and Ore Deposits of the Takilma and Waldo Mining Districts - Unavailable since the 1933, this publication was originally compiled by the United States Geological Survey and includes details on gold and copper mining in the Takilma and Waldo Districts of Josephine County, Oregon. The Waldo and Takilma mining districts are most notable for the fact that the earliest large scale mining of placer gold and copper in Oregon took place in these two areas. Included in this report are details about some of the earliest large gold mines in the state such as the Llano de Oro, High Gravel, Cameron, Platerica, Deep Gravel and others, as well as copper mines such as the famous Queen of Bronze mine, the Waldo, Lily and Cowboy mines. In addition to geological examinations, insights are also provided into the production, day to day operations and early histories of these mines, as well as calculations of known mineral reserves in the area. This volume also includes six maps and 20 original illustrations. **8.5" X 11", 74 ppgs. Retail Price: $9.99**

Gold Mines of Oregon - Oregon mining historian Kerby Jackson introduces us to a classic work on Oregon's mining history in this important re-issue of Bulletin 61, otherwise known as "Gold and Silver In Oregon". Unavailable since 1968, this important publication was originally compiled by geologists Howard C. Brooks and Len Ramp of the Oregon Department of Geology and Mineral Industries and includes detailed descriptions, histories and the geology of over 450 gold mines Oregon. Included are notes on the history, geology and gold production statistics of all the major mining areas in Oregon including the Klamath Mountains, the Blue Mountains and the North Cascades. While gold is where you find it, as every miner knows, the path to success is to prospect for gold where it was previously found. 8.5" X 11", 344 ppgs. Retail Price: $24.99

Mines and Mineral Resources of Curry County Oregon - Originally published in 1916, this important publication on Oregon Mining has not been available for nearly a century. Included are rare insights into the history, production and locations of dozens of gold mines in Curry County, Oregon, as well as detailed information on important Oregon mining districts in that area such as those at Agness, Bald Face Creek, Mule Creek, Boulder Creek, China Diggings, Collier Creek, Elk River, Gold Beach, Rock Creek, Sixes River and elsewhere. Particular attention is especially paid to the famous beach gold deposits of this portion of the Oregon Coast. 8.5" X 11", 140 ppgs. Retail Price: $11.99

Chromite Mining in South West Oregon - Originally published in 1961, this important publication on Oregon Mining has not been available for nearly a century. Included are rare insights into the history, production and locations of nearly 300 chromite mines in South Western Oregon. 8.5" X 11", 184 ppgs. Retail Price: $14.99

Mineral Resources of Douglas County Oregon - Originally published in 1972, this important publication on Oregon Mining has not been available for nearly forty years. Included are rare insights into the geology, history, production and locations of numerous gold mines and other mining properties in Douglas County, Oregon. 8.5" X 11", 124 ppgs. Retail Price: $11.99

Mineral Resources of Coos County Oregon - Originally published in 1972, this important publication on Oregon Mining has not been available for nearly forty years. Included are rare insights into the geology, history, production and locations of numerous gold mines and other mining properties in Coos County, Oregon. 8.5" X 11", 100 ppgs. Retail Price: $11.99

Mineral Resources of Lane County Oregon - Originally published in 1938, this important publication on Oregon Mining has not been available for nearly seventy five years. Included are extremely rare insights into the geology and mines of Lane County, Oregon, in particular in the Bohemia, Blue River, Oakridge, Black Butte and Winberry Mining Districts. 8.5" X 11", 82 ppgs. Retail Price: $9.99

Mineral Resources of the Upper Chetco River of Oregon: Including the Kalmiopsis Wilderness - Originally published in 1975, this important publication on Oregon Mining has not been available for nearly forty years. Withdrawn under the 1872 Mining Act since 1984, real insight into the minerals resources and mines of the Upper Chetco River has long been unavailable due to the remoteness of the area. Despite this, the decades of battle between property owners and environmental extremists over the last private mining inholding in the area has continued to pique the interest of those interested in mining and other forms of natural resource use. Gold mining began in the area in the 1850's and has a rich history in this geographic area, even if the facts surrounding it are little known. Included are twenty two rare photographs, as well as insights into the Becca and Morning Mine, the Emmly Mine (also known as Emily Camp), the Frazier Mine, the Golden Dream or Higgins Mine, Hustis Mine, Peck Mine and others. 8.5" X 11", 64 ppgs. Retail Price: $8.99

Gold Dredging in Oregon - Originally published in 1939, this important publication on Oregon Mining has not been available for nearly seventy five years. Included are extremely rare insights into the history and day to day operations of the dragline and bucketline gold dredges that once worked the placer gold fields of South West and North East Oregon in decades gone by. Also included are details into the areas that were worked by gold dredges in Josephine, Jackson, Baker and Grant counties, as well as the economic factors that impacted this mining method. This volume also offers a unique look into the values of river bottom land in relation to both farming and mining, in how farm lands were mined, re-soiled and reclamated after the dredges worked them. Featured are hard to find maps of the gold dredge fields, as well as rare photographs from a bygone era. 8.5" X 11", 86 ppgs. Retail Price: $8.99

Quick Silver Mining in Oregon - Originally published in 1963, this important publication on Oregon Mining has not been available for over fifty years. This publication includes details into the history and production of Elemental Mercury or Quicksilver in the State of Oregon. 8.5" X 11", 238 ppgs. Retail Price: $15.99

Mines of the Greenhorn Mining District of Grant County Oregon - Originally published in 1948, this important publication on Oregon Mining has not been available for over sixty five years. In this publication are rare insights into the mines of the famous Greenhorn Mining District of Grant County, Oregon, especially the famous Morning Mine. Also included are details on the Tempest, Tiger, Bi-Metallic, Windsor, Psyche, Big Johnny, Snow Creek, Banzette and Paramount Mines, as well as prospects in the vicinities in the famous mining areas of Mormon Basin, Vinegar Basin and Desolation Creek. Included are hard to find mine maps and dozens of rare photographs from the bygone era of Grant County's rich mining history. 8.5" X 11", 72 ppgs. Retail Price: $9.99

Geology of the Wallowa Mountains of Oregon: Part I (Volume 1) - Originally published in 1938, this important publication on Oregon Mining has not been available for nearly seventy five years. Included are details on the geology of this unique portion of North Eastern Oregon. This is the first part of a two book series on the area. Accompanying the text are rare photographs and historic maps. **8.5" X 11", 92 ppgs. Retail Price: $9.99**

Geology of the Wallowa Mountains of Oregon: Part II (Volume 2) - Originally published in 1938, this important publication on Oregon Mining has not been available for nearly seventy five years. Included are details on the geology of this unique portion of North Eastern Oregon. This is the first part of a two book series on the area. Accompanying the text are rare photographs and historic maps. **8.5" X 11", 94 ppgs. Retail Price: $9.99**

Field Identification of Minerals For Oregon Prospectors - Originally published in 1940, this important publication on Oregon Mining has not been available for nearly seventy five years. Included in this volume is an easy system for testing and identifying a wide range of minerals that might be found by prospectors, geologists and rockhounds in the State of Oregon, as well as in other locales. Topics include how to put together your own field testing kit and how to conduct rudimentary tests in the field. This volume is written in a clear and concise way to make it useful even for beginners. **8.5" X 11", 158 ppgs. Retail Price: $14.99**

The Bohemia Mining District of Oregon - Originally published in 1900, this important publication on Oregon Mining has not been available for over a century. Included in this volume are important insights into the famous Bohemia Mining District of Oregon, including the histories and locations of important gold mines in the area such as the Ophir Mine, Clarence, Acturas, Peek-a-boo, White Swan, Combination Mine, the Musick Mine, The California, White Ghost, The Mystery, Wall Street, Vesuvius, Story, Lizzie Bullock, Delta, Elsie Dora, Golden Slipper, Broadway, Champion Mine, Knott, Noonday, Helena, White Wings, Riverside and others. Also included are notes on the nearby Blue River Mining District. **8.5" X 11", 58 ppgs. Retail Price: $9.99**

The Gold Fields of Eastern Oregon - Unavailable since 1900, this publication was originally compiled by the Baker City Chamber of Commerce Offering important insights into the gold mining history of Eastern Oregon, "The Gold Fields of Eastern Oregon" sheds a rare light on many of the gold mines that were operating at the turn of the 19th Century in Baker County and Grant County in North Eastern Oregon. Some of the areas featured include the Cable Cove District, Baisely-Elhorn, Granite, Red Boy, Bonanza, Susanville, Sparta, Virtue, Vaughn, Sumpter, Burnt River, Rye Valley and other mining districts. Included is basic information on not only many gold mines that are well known to those interested in Eastern Oregon mining history, but also many mines and prospects which have been mostly lost to the passage of time. Accompanying are numerous rare photos **8.5" X 11", 78 ppgs. Retail Price: $10.99**

Gold Mining in Eastern Oregon - Originally published·in 1938, this important publication on Oregon Mining has not been available for over a century. Included in this volume are important insights into the famous mining districts of Eastern Oregon during the late 1930's. Particular attention is given to those gold mines with milling and concentrating facilities in the Greenhorn, Red Boy, Alamo, Bonanza, Granite, Cable Cove, Cracker Creek, Virtue, Keating, Medical Springs, Sanger, Sparta, Chicken Creek, Mormon Basin, Connor Creek, Cornucopia and the Bull Run Mining Districts. Some of the mines featured include the Ben Harrison, North Pole-Columbia, Highland Maxwell, Baisley-Elkhorn, White Swan, Balm Creek, Twin Baby, Gem of Sparta, New Deal, Gleason, Gifford-Johnson, Cornucopia, Record, Bull Run, Orion and others. Of particular interest are the mill flow sheets and descriptions of milling operations of these mines. **8.5" X 11", 68 ppgs. Retail Price: $8.99**

The Gold Belt of the Blue Mountains of Oregon - Originally published in 1901, this important publication on Oregon Mining has not been available for over a century. Included in this volume are rare insights into the gold deposits of the Blue Mountains of North East Oregon, including the history of their early discovery and early production. Extensive details are offered on this important mining area's mineralogy and economic geology, as well as insights into nearby gold placers, silver deposits and copper deposits. Featured are the Elkhorn and Rock Creek mining districts, the Pocahontas district, Auburn and Minersville districts, Sumpter and Cracker Creek, Cable Cove, the Camp Carson district, Granite, Alamo, Greenhorn, Robinsonville, the Upper Burnt River Valley and Bonanza districts, Susanville, Quartzburg, Canyon Creek, Virtue, the Copper Butte district, the North Powder River, Sparta, Eagle Creek, Cornucopia, Pine Creek, Lower Powder River, the Upper Snake River Canyon, Rye Valley, Lower Burnt River Valley, Mormon Basin, the Malheur and Clarks Creek districts, Sutton Creek and others. Of particular interest are important details on numerous gold mines and prospects in these mining districts, including their locations, histories, geology and other important information, as well as information on silver, copper and fire opal deposits. **8.5" X 11", 250 ppgs. Retail Price: $24.99**

Mining in the Cascades Range of Oregon - Originally published in 1938, this important publication on Oregon Mining has not been available for over seventy five years. Included in this volume are rare insights into the gold mines and other types of metal mines in the Cascades Mountain Range of Oregon. Some of the important mining areas covered include the famous Bohemia Mining District, the North Santiam Mining District, Quartzville Mining District, Blue River Mining District, Fall Creek Mining District, Oakridge District, Zinc District, Buzzard-Al Sarena District, Grand Cove, Climax District and Barron Mining District. Of particular interest are important details on over 100 mines and prospects in these mining districts, including their locations, histories, geology and other important information. 8.5" X 11", 170 ppgs. Retail Price: $14.99

Beach Gold Placers of the Oregon Coast - Originally published in 1934, this important publication on Oregon Mining has not been available for over 80 years. Included in this volume are rare insights into the beach gold deposits of the State of Oregon, including their locations, occurance, composition and geology. Of particular interest is information on placer platinum in Oregon's rich beach deposits. Also included are the locations and other information on some famous Oregon beach mines, including the Pioneer, Eagle, Chickamin, Iowa and beach placer mines north of the mouth of the Rogue River. 8.5" X 11", 60 ppgs. Retail Price: $8.99

Mineralogical Composition of the Sands of the Oregon Coast: From Coos Bay to the Columbia - Published in 1945, he text features hard to find information on the composition of the gold bearing black sands of the South West Oregon Coast, offering a unique insight to prospectors in search of Oregon's legendary beach gold. 104 ppgs, $9.99

Manganese Mining in Oregon - First released in 1942 and now out of print, this special reprint edition of "Manganese in Oregon" was originally published by the Oregon Department of Geology and Mineral Industries. The text features hard to find information on the mining of Manganese in Oregon, including details and maps of Oregon manganese mines and prospects. 108 ppgs, 9.99

Medford Oregon As A Mining Center - Written in 1912, this hard to find publication includes valuable insights into the mining history of South West Oregon. This small book contains interesting information on the gold, copper and mining industry in Southern Oregon as it existed just prior to World War One, shedding light on some of the important mines in the area. Included are rare photographs and vintage advertising of the day. 80 ppgs, 9.99

Mineral Resources of Curry County Oregon - First released in 1977 and now out of print, this special reprint edition of "Geology, Mineral Resources and Rock Materials of Curry County, Oregon" was originally published in cooperation of Curry County, Oregon and the Oregon Department of Geology and Mineral Industries. The text features hard to find information on not only the mining of gold and other metals in Curry County, but also aggregate mining in the area. 102 ppgs, 11.99

Origin of the Gold Bearing Black Sands of the Coast of South West Oregon - First released in 1943 and now out of print, this special reprint edition of "The Origin of the Black Sands of the South West Oregon Coast" was originally published by the Oregon Department of Geology and Mineral Industries. The text features hard to find information on the origin of the gold bearing black sands of the South West Oregon Coast, offering a unique insight to prospectors in search of Oregon's legendary beach gold. 52 ppgs, 8.99

South West Oregon Mining - Leading mining historian Kerby Jackson introduces us to six classic small mining publications on the Gold Mining Industry in Southern Oregon. This small book consists of a compilation of USGS J.S. Diller's "Mines of the Riddles Quadrangle", "The Rogue River Valley Coal Fields" and "Mineral Resources of the Grants Pass Quadrangle", the Grants Pass Commercial Club's rare publication "Mining in Josephine County, Oregon" and the USGS publication "The Distribution of Placer Gold in the Sixes River, South West Oregon". Also included is F.W. Libbey's legendary article on the Southern Oregon Mining Industry, "Lest We Forget", which appeared in the publication of the Oregon State Department of Geology and Mineral Industries in the early 1960's. This compilation offers a unique perspective on mining in South West Oregon and includes considerable information on mines in Josephine, Jackson and Coos Counties. 142 ppgs, 14.99

Geology and Mineral Resources of the Gasquet Quadrangle of California-Oregon - First published in 1953, it has been unavailable for over a century and sheds important light on the geological features and mineral resources of this portion of Northern California and Southern Oregon. 80 ppgs, 9.99

Idaho Mining Books

Gold in Idaho - Unavailable since the 1940's, this publication was originally compiled by the Idaho Bureau of Mines and includes details on gold mining in Idaho. Included is not only raw data on gold production in Idaho, but also valuable insight into where gold may be found in Idaho, as well as practical information on the gold bearing rocks and other geological features that will assist those looking for placer and lode gold in the State of Idaho. This volume also includes thirteen gold maps that greatly enhance the practical usability of the information contained in this small book detailing where to find gold in Idaho. **8.5" X 11", 72 ppgs. Retail Price: $9.99**

Geology of the Couer D'Alene Mining District of Idaho - Unavailable since 1961, this publication was originally compiled by the Idaho Bureau of Mines and Geology and includes details on the mining of gold, silver and other minerals in the famous Coeur D'Alene Mining District in Northern Idaho. Included are details on the early history of the Coeur D'Alene Mining District, local tectonic settings, ore deposit features, information on the mineral belts of the Osburn Fault, as well as detailed information on the famous Bunker Hill Mine, the Dayrock Mine, Galena Mine, Lucky Friday Mine and the infamous Sunshine Mine. This volume also includes sixteen hard to find maps. **8.5" X 11", 70 ppgs. Retail Price: $9.99**

The Gold Camps and Silver Cities of Idaho - Originally published in 1963, this important publication on Idaho Mining has not been available for nearly fifty years. Included are rare insights into the history of Idaho's Gold Rush, as well as the mad craze for silver in the Idaho Panhandle. Documented in fine detail are the early mining excitements at Boise Basin, at South Boise, in the Owyhees, at Deadwood, Long Valley, Stanley Basin and Robinson Bar, at Atlanta, on the famous Boise River, Volcano, Little Smokey, Banner, Boise Ridge, Hailey, Leesburg, Lemhi, Pearl, at South Mountain, Shoup and Ulysses, Yellow Jacket and Loon Creek. The story follows with the appearance of Chinese miners at the new mining camps on the Snake River, Black Pine, Yankee Fork, Bay Horse, Clayton, Heath, Seven Devils, Gibbonsville, Vienna and Sawtooth City. Also included are special sections on the Idaho Lead and Silver mines of the late 1800's, as well as the mining discoveries of the early 1900's that paved the way for Idaho's modern mining and mineral industry. Lavishly illustrated with rare historic photos, this volume provides a one of a kind documentary into Idaho's mining history that is sure to be enjoyed by not only modern miners and prospectors who still scour the hills in search of nature's treasures, but also those enjoy history and tromping through overgrown ghost towns and long abandoned mining camps. **8.5" X 11", 186 ppgs. Retail Price: $14.99**

Ore Deposits and Mining in North Western Custer County Idaho - Unavailable since 1913, this important publication was originally published by the Us Department of the Interior and has been unavailable for a century. Included are fine details on the geology, geography, gold placers and gold and silver bearing quartz veins of the mining region of North West Custer County, Idaho. Of particular interest is a rare look at the mines and prospects of the region, including those such as the Ramshorn Mine, SkyLark, Riverview, Excelsior, Beardsley, Pacific, Hoosier, Silver Brick, Forest Rose and dozens of others in the Bay Horse Mining District. Also covered are the mines of the Yankee Fork District such as the Lucky Boy, Badger, Black, Enterprise, Charles Dickens, Morrison, Golden Sunbeam, Montana, Golden Gate and others, as well as those in the Loon Mining District. **8.5" X 11", 126 ppgs. Retail Price: $12.99**

Gold Rush To Idaho - Unavailable since 1963, this important publication was originally published by the Idaho Bureau of Mines and has been unavailable for 50 years. "Gold Rush To Idaho" revisits the earliest years of the discovery of gold in Idaho Territory and introduces us to the conditions that the pioneer gold seekers met when they blazed a trail through the wilderness of Idaho's mountains and discovered the precious yellow metal at Oro Fino and Pierce. Subsequent rushes followed at places like Elk City, Newsome, Clearwater Station, Florence, Warrens and elsewhere. Of particular interest is a rare look at the hardships that the first miners in Idaho met with during their day to day existences and their attempts to bring law and order to their mining camps. **8.5" X 11", 88 ppgs. Retail Price: $9.99**

The Geology and Mines of Northern Idaho and North Western Montana - Unavailable since 1909, this important publication was originally published by the Us Department of the Interior and has been unavailable for a century. Included are fine details on the geology and geography of the mining regions of Northern Idaho and North Western Montana. Of particular interest is a rare look at the mines and prospects of the region, including those in the Pine Creek Mining District, Lake Pend Oreille district, Troy Mining District, Sylvanite District, Cabinet Mining District, Prospect Mining District and the Missoula Valley. Some of the mines featured include the Iron Mountain, Silver Butte, Snowshoe, Grouse Mountain Mine and others. **8.5" X 11", 142 ppgs. Retail Price: $12.99**

Mining in the Alturas Quadrangle of Blaine County Idaho - Unavailable since 1922, this important publication was originally published by the Idaho Bureau of Mines and has been unavailable for ninety years. Topics include the geology, rock formations and the formation of ore deposits in this important mining area of Idaho. Of particular focus is information on the local geology, quartz veins and ore deposits of this portion of Idaho. Included are hard to find details, including the descriptions and locations of numerous gold and silver mines in the area including the Silver King, Pilgrim, Columbia, Lone Jack, Sunbeam, Pride of the West, Lucky Boy, Scotia, Atlanta, Beaver-Bidwell and others mines and prospects. **8.5" X 11", 56 ppgs. Retail Price: $8.99**

Mining in Lemhi County Idaho - Originally published in 1913, this important book on Idaho Mining has not been available to miners for over a century. Included are rare insights into hundreds of gold, silver, copper and other mines in this famous Idaho mining area. Details include the locations, geology, history, production and other facts of the mines of this region, not only gold and silver hardrock mines, but also gold placer mines, lead-silver deposits, copper mines, cobalt-nickel deposits, tungsten and tin mines . It is lavishly illustrated with hard to find photos of the period and rare mining maps. Some of the vicinities featured include the Nicholia Mining District, Spring Mountain District, Texas District, Blue Wing District, Junction District, McDevitt District, Pratt Creek, Eldorado District, Kirtley Creek, Carmen Creek, Gibbonsville, Indian Creek, Mineral Hill District, Mackinaw, Eureka District, Blackbird District, YellowJacket District, Gravel Range District, Junction District, Parker Mountain and other mining districts. **8.5" X 11", 226 ppgs. Retail Price: $19.99**

Mining in Shoshone County Idaho - First published in 1923, it has been unavailable for over a century and sheds important light on the mining history of Shoshone County, Idaho. Some of the topics include the history of mining in Shoshone County, a look at the local geology and ore characteristics of lead-silver deposits, zinc deposits, copper, antimony, gold and other minerals. Also included are insights into the history, production, characteristics and locations of numerous mines in the area. 198 ppgs, 15.99

Utah Mining Books

Fluorite in Utah - Unavailable since 1954, this publication was originally compiled by the USGS, State of Utah and U.S. Atomic Energy Commission and details the mining of fluorspar, also known as fluorite in the State of Utah. Included are details on the geology and history of fluorspar (fluorite) mining in Utah, including details on where this unique gem mineral may be found in the State of Utah. **8.5" X 11", 60 ppgs. Retail Price: $8.99**

The Gold Hill Mining District of Utah - First published in 1935, it has been unavailable since those days and sheds important light on the mines, history and geology of Utah's Gold Hill Mining District. Included are rare insights into this important mining area, including the locations, histories and details of numerous mines. This volume is well illustrated with geological diagrams, as well as hard to find maps of some of the most important mines in this district. 202 ppgs., 19.99

The Mines, Miners and Minerals of Utah - First published in 1896, it has been unavailable since those days and sheds important light on the early mines and miners of Pioneer Utah, as well as the minerals which they won from the earth by laborious hard physical labor and sheer determination. Included are rare insights into the early mining history of Utah, as well details on hundreds of gold, silver and copper mines. 376 ppgs., 24.99

California Mining Books

The Tertiary Gravels of the Sierra Nevada of California - Mining historian Kerby Jackson introduces us to a classic mining work by Waldemar Lindgren in this important re-issue of The Tertiary Gravels of the Sierra Nevada of California. Unavailable since 1911, this publication includes details on the gold bearing ancient river channels of the famous Sierra Nevada region of California. **8.5" X 11", 282 ppgs. Retail Price: $19.99**

The Mother Lode Mining Region of California - Unavailable since 1900, this publication includes details on the gold mines of California's famous Mother Lode gold mining area. Included are details on the geology, history and important gold mines of the region, as well as insights into historic mining methods, mine timbering, mining machinery, mining bell signals and other details on how these mines operated. Also included are insights into the gold mines of the California Mother Lode that were in operation during the first sixty years of California's mining history. **8.5" X 11", 176 ppgs. Retail Price: $14.99**

Lode Gold of the Klamath Mountains of Northern California and South West Oregon - Unavailable since 1971, this publication was originally compiled by Preston E. Hotz and includes details on the lode mining districts of Oregon and California's Klamath Mountains. Included are details on the geology, history and important lode mines of the French Gulch, Deadwood, Whiskeytown, Shasta, Redding, Muletown, South Fork, Old Diggings, Dog Creek (Delta), Bully Choop (Indian Creek), Harrison Gulch, Hayfork, Minersville, Trinity Center, Canyon Creek, East Fork, New River, Denny, Liberty (Black Bear), Cecilville, Callahan, Yreka, Fort Jones and Happy Camp mining districts in California, as well as the Ashland, Rogue River, Applegate, Illinois River, Takilma, Greenback, Galice, Silver Peak, Myrtle Creek and Mule Creek districts of South Western Oregon. Also included are insights into the mineralization and other characteristics of this important mining region. **8.5" X 11", 100 ppgs. Retail Price: $10.99**

Mines and Mineral Resources of Shasta County, Siskiyou County, Trinity County: California - Unavailable since 1915, this publication was originally compiled by the California State Mining Bureau and includes details on the gold mines of this area of Northern California. Also included are insights into the mineralization and other characteristics of this important mining region, as well as the location of historic gold mines. **8.5" X 11", 204 ppgs. Retail Price: $19.99**

Geology of the Yreka Quadrangle, Siskiyou County, California - Unavailable since 1977, this publication was originally compiled by Preston E. Hotz and includes details on the geology of the Yreka Quadrangle of Siskiyou County, California. Also included are insights into the mineralization and other characteristics of this important mining region. 8.5" X 11", 78 ppgs. Retail Price: $7.99

Mines of San Diego and Imperial Counties, California - Originally published in 1914, this important publication on California Mining has not been available for a century. This publication includes important information on the early gold mines of San Diego and Imperial County, which were some of the first gold fields mined in California by early Spanish and Mexican miners before the 49ers came on the scene. Included are not only details on early mining methods in the area, production statistics and geological information, but also the location of the early gold mines that helped make California "The Golden State". Also included are details on the mining of other minerals such as silver, lead, zinc, manganese, tungsten, vanadium, asbestos, barite, borax, cement, clay, dolomite, fluospar, gem stones, graphite, marble, salines, petroleum, stronium, talc and others. 8.5" X 11", 116 ppgs. Retail Price: $12.99

Mines of Sierra County, California - Unavailable since 1920, this publication was originally compiled by the California State Mining Bureau and includes details on the gold mines of Sierra County, California. Also included are insights into the mineralization and other characteristics of this important mining region, as well as the location of historic gold mines. 8.5" X 11", 156 ppgs. Retail Price: $19.99

Mines of Plumas County, California - Unavailable since 1918, this publication was originally compiled by the California State Mining Bureau and includes details on the gold mines of Plumas County, California. Also included are insights into the mineralization and other characteristics of this important mining region, as well as the location of historic gold mines. 8.5" X 11", 200 ppgs. Retail Price: $19.99

Mines of El Dorado, Placer, Sacramento and Yuba Counties, California - Originally published in 1917, this important publication on California Mining has not been available for nearly a century. This publication includes important information on the early gold mines of El Dorado County, Placer County, Sacramento County and Yuba County, which were some of the first gold fields mined by the Forty-Niners during the California Gold Rush. Included are not only details on early mining methods in the area, production statistics and geological information, but also the location of the early gold mines that helped make California "The Golden State". Also included are insights into the early mining of chrome, copper and other minerals in this important mining area. 8.5" X 11", 204 ppgs. Retail Price: $19.99

Mines of Los Angeles, Orange and Riverside Counties, California - Originally published in 1917, this important publication on California Mining has not been available for nearly a century. This publication includes important information on the early gold mines of Los Angeles County, Orange County and Riverside County, which were some of the first gold fields mined in California by early Spanish and Mexican miners before the 49ers came on the scene. Included are not only details on early mining methods in the area, production statistics and geological information, but also the location of the early gold mines that helped make California "The Golden State". 8.5" X 11", 146 ppgs. Retail Price: $12.99

Mines of San Bernadino and Tulare Counties, California - Originally published in 1917, this important publication on California Mining has not been available for nearly a century. This publication includes important information on the early gold mines of San Bernadino and Tulare County, which were some of the first gold fields mined in California by early Spanish and Mexican miners before the 49ers came on the scene. Included are not only details on early mining methods in the area, production statistics and geological information, but also the location of the early gold mines that helped make California "The Golden State". Also included are details on the mining of other minerals such as copper, iron, lead, zinc, manganese, tungsten, vanadium, asbestos, barite, borax, cement, clay, dolomite, fluospar, gem stones, graphite, marble, salines, petroleum, stronium, talc and others. 8.5" X 11", 200 ppgs. Retail Price: $19.99

Chromite Mining in The Klamath Mountains of California and Oregon - Unavailable since 1919, this publication was originally compiled by J.S. Diller of the United States Department of Geological Survey and includes details on the chromite mines of this area of Northern California and Southern Oregon. Also included are insights into the mineralization and other characteristics of this important mining region, as well as the location of historic mines. Also included are insights into chromite mining in Eastern Oregon and Montana. 8.5" X 11", 98 ppgs. Retail Price: $9.99

Mines and Mining in Amador, Calaveras and Tuolumne Counties, California - Unavailable since 1915, this publication was originally compiled by William Tucker and includes details on the mines and mineral resources of this important California mining area. Included are details on the geology, history and important gold mines of the region, as well as insights into other local mineral resources such as asbestos, clay, copper, talc, limestone and others. Also included are insights into the mineralization and other characteristics of this important portion of California's Mother Lode mining region. 8.5" X 11", 198 ppgs. Retail Price: $14.99

The Cerro Gordo Mining District of Inyo County California - Unavailable since 1963, this publication was originally compiled by the United States Department of Interior. Included are insights into the mineralization and other characteristics of this important mining region of Southern California. Topics include the mining of gold and silver in this important mining district in Inyo County, California, including details on the history, production and locations of the Cerro Gordo Mine, the Morning Star Mine, Estelle Tunnel, Charles Lease Tunnel, Ignacio, Hart, Crosscut Tunnel, Sunset, Upper Newtown, Newtown, Ella, Perseverance, Newsboy, Belmont and other silver and gold mines in the Cerro Gordo Mining District. This volume also includes important insights into the fossil record, geologic formations, faults and other aspects of economic geology in this California mining district. 8.5″ X 11″, 104 ppgs. Retail Price: $10.99

Mining in Butte, Lassen, Modoc, Sutter and Tehama Counties of California - Unavailable since 1917, this publication was originally compiled by the United States Department of Interior. Included are insights into the mineralization and other characteristics of this important mining region of California. Topics include the mining of asbestos, chromite, gold, diamonds and manganese in Butte County, the mining of gold and copper in the Hayden Hill and Diamond Mountain mining districts of Lassen County, the mining of coal, salt, copper and gold in the High Grade and Winters mining districts of Modoc County, gold mining in Sutter County and the mining of gold, chromite, manganese and copper in Tehama County. This volume also includes the production records and locations of numerous mines in this important mining region. 8.5″ X 11″, 114 ppgs. Retail Price: $11.99

Mines of Trinity County California - Originally published in 1965, this important publication on California Mining has not been available for nearly fifty years. This publication includes important information on mines and mining in Trinity County, California, as well insights into the mineralization and geology of this important mining area in Northern California. Included are extensive details on hardrock and placer gold mines and prospects, including charts showing the locations of these historic mines.. 8.5″ X 11″, 144 ppgs. Retail Price: $12.99

Mines of Kern County California - Originally published in 1962, this important publication on California Mining has not been available for nearly fifty years. This publication includes important information on mines and mining in Kern County, California, as well insights into the mineralization and geology of this important mining area in California. Included are extensive details on hardrock and placer gold mines and prospects, including charts showing the locations of these historic mines. 8.5″ X 11″, 398 ppgs. Retail Price: $24.99

Mines of Calaveras County California - Originally published in 1962, this important publication on California Mining has not been available for nearly fifty years. This publication includes important information on mines and mining in Calaveras County, California, as well insights into the mineralization and geology of this important mining area in Northern California. Included are extensive details on hardrock and placer gold mines and prospects, including charts showing the locations of these historic mines. 8.5″ X 11″, 236 ppgs. Retail Price: $19.99

Lode Gold Mining in Grass Valley California - Unavailable since 1940, this publication was originally compiled by the United States Department of Interior. Included are insights into the gold mineralization and other characteristics of this important mining region of Nevada County, California. This volume also includes important insights into the geologic formations, faults and other aspects of economic geology in this California mining district. Of particular interest are the fine details on many hardrock gold mines in the area, including their locations, histories, development and mineralization. Some of the mines featured include the Gold Hill Mine, Massachusetts Hill, Boundary, Peabody, Golden Center, North Star, Omaha, Lone Jack, Homeward Bound, Hartery, Wisconsin, Allison Ranch, Phoenix, Kate Hayes, W.Y.O.D., Empire, Rich Hill, Daisy Hill, Orleans, Sultana, Centennial, Conlin, Ben Franklin, Crown Point and many others. 8.5″ X 11″, 148 ppgs. Retail Price: $12.99

Lode Mining in the Alleghany District of Sierra County California - Unavailable since 1913, this publication was originally compiled by the United States Department of Interior. Included are insights into the mineralization and other characteristics of this important mining region of Sierra County. Included are details on the history, production and locations of numerous hardrock gold mines in this famous California area, including the Tightner Mine, Minnie D., Osceola, Eldorado, Twenty One, Sherman, Kenton, Oriental, Rainbow, Plumbago, Irelan, Gold Canyon, North Fork, Federal, Kate Hardy and others. This volume also includes important insights into the fossil record, geologic formations, faults and other aspects of economic geology in this California mining district. 8.5″ X 11″, 48 ppgs. Retail Price: $7.99

Six Months In The Gold Mines During The California Gold Rush - Unavailable since 1850, this important work is a first hand account of one "49'ers" personal experience during the great California Gold Rush, shedding important light on one of the most exciting periods in the history of not only California, but also the world. Compiled from journals written between 1847 and 1849 by E. Gould Buffum, a native of New York, "Six Months In The Gold Mines During The California Gold Rush" offers a rare look into the day to day lives of the people who came to California to work in her gold mines when the state was still a great frontier. 8.5″ X 11″, 290 ppgs. Retail Price: $19.99

Quartz Mines of the Grass Valley Mining District of California - Unavailable since 1867, this important publication has not been available since those days. This rare publication offers a short dissertation on the early hardrock mines in this important mining district in the California Mother Lode region between the 1850's and 1860's. Also included are hard to find details on the mineralization and locations of these mines, as well as how they were operated in those day. **8.5" X 11", 44 ppgs. Retail Price: $8.99**

Gold Rush on the Feather River - **First published in 1924, this short publication by G.C. Mansfield sheds important light on the early history of gold mining on the Feather River. Included are rare insights into the first decade of gold mining and the early mining camps of the Feather River during the 1850's. 64 ppgs., 9.99**

The Bodie Mining District of California - First published in 1986, it has been unavailable since those days and sheds important light on this famous mining area. Included are the history, characteristics and locations of numerous old mines around the ghost town of Bodie.
64 ppgs, 8.99

Geology and Mineral Resources of the Gasquet Quadrangle of California-Oregon - First published in 1953, it has been unavailable for over a century and sheds important light on the geological features and mineral resources of this portion of Northern California and Southern Oregon.
80 ppgs, 9.99

Alaska Mining Books

Ore Deposits of the Willow Creek Mining District, Alaska - Unavailable since 1954, this hard to find publication includes valuable insights into the Willow Creek Mining District near Hatcher Pass in Alaska. The publication includes insights into the history, geology and locations of the well known mines in the area, including the Gold Cord, Independence, Fern, Mabel, Lonesome, Snowbird, Schroff-O'Neil, High Grade, Marion Twin, Thorpe, Webfoot, Kelly-Willow, Lane, Holland and others. **8.5" X 11", 96 ppgs. Retail Price: $9.99**

The Juneau Gold Belt of Alaska - Unavailable since 1906, this hard to find publication includes valuable insights into the gold mines around Juneau, Alaska. The publication includes important details into the history, geology and locations of the well known gold mines and prospects in the area, including those around Windham Bay, Holkham Bay, Port Snettisham, on Grindstone and Rhine Creeks, Gold Creek, Douglas Island, Salmon Creek, Lemon Creek, Nugget Creek, from the Mendenhall River to Berners Bay, McGinnis Creek, Montana Creek, Peterson Creek, Windfall Creek, the Eagle River, Yankee Basin, Yankee Curve, Kowee Creek and elsewhere. Not only are gold placer mines included, but also hardrock gold mines. **8.5" X 11", 224 ppgs. Retail Price: $19.99**

Mining in the Jumbo Basin of Alaska - Unavailable since 1953, this hard to find publication includes valuable insights into the mines and geology of the Jumbo Basin. The publication includes important details into the history, geology and locations of the well known gold mines and prospects in the famous Jumbo Basin Mining Region of Alaska.
72 ppgs, 9.99

The Rampart Placer Gold Region of Alaska - Unavailable since 1906, this hard to find publication includes valuable insights into the placer gold mines of the Rampart Mining Region. The publication includes important details into the history, geology and locations of the well known gold mines and prospects in the famous Rampart Mining Region of Alaska.
78 ppgs, 10.99

Arizona Mining Books

Mines and Mining in Northern Yuma County Arizona - Originally published in 1911, this important publication on Arizona Mining has not been available for over a hundred years. Included are rare insights into the gold, silver, copper and quicksilver mines of Yuma County, Arizona together with hard to find maps and photographs. Some of the mines and mining districts featured include the Planet Copper Mine, Mineral Hill, the Clara Consolidated Mine, Viati Mine, Copper Basin prospect, Bowman Mine, Quartz King, Billy Mack, Carnation, the Wardwell and Osbourne, Valensuella Copper, the Mariquita, Colonial Mine, the French American, the New York-Plomosa, Guadalupe, Lead Camp, Mudersbach Copper Camp, Yellow Bird, the Arizona Northern (Salome Strike), Bonanza (Harqua Hala), Golden Eagle, Hercules, Socorro and others. **8.5" X 11", 144 ppgs. Retail Price: $11.99**

The Aravaipa and Stanley Mining Districts of Graham County Arizona - Originally published in 1925, this important publication on Arizona Mining has not been available for nearly ninety years. Included are rare insights into the gold and silver mines of these two important mining districts, together with hard to find maps. **8.5" X 11", 140 ppgs. Retail Price: $11.99**

<u>Gold in the Gold Basin and Lost Basin Mining Districts of Mohave County, Arizona</u> - This volume contains rare insights into the geology and gold mineralization of the Gold Basin and Lost Basin Mining Districts of Mohave County, Arizona that will be of benefit to miners and prospectors. Also included is a significant body of information on the gold mines and prospects of this portion of Arizona. This volume is lavishly illustrated with rare photos and mining maps. **8.5" X 11", 188 ppgs. Retail Price: $19.99**

<u>Mines of the Jerome and Bradshaw Mountains of Arizona</u> - This important publication on Arizona Mining has not been available for ninety years. This volume contains rare insights into the geology and ore deposits of the Jerome and Bradshaw Mountains of Arizona that will be of benefit to miners and prospectors who work those areas. Included is a significant body of information on the mines and prospects of the Verde, Black Hills, Cherry Creek, Prescott, Walker, Groom Creek, Hassayampa, Bigbug, Turkey Creek, Agua Fria, Black Canyon, Peck, Tiger, Pine Grove, Bradshaw, Tintop, Humbug and Castle Creek Mining Districts. This volume is lavishly illustrated with rare photos and mining maps. **8.5" X 11", 218 ppgs. Retail Price: $19.99**

<u>The Ajo Mining District of Pima County Arizona</u> - This important publication on Arizona Mining has not been available for nearly seventy years. This volume contains rare insights into the geology and mineralization of the Ajo Mining District in Pima County, Arizona and in particular the famous New Cornelia Mine. **8.5" X 11", 126 ppgs. Retail Price: $11.99**

<u>Mining in the Santa Rita and Patagonia Mountains of Arizona</u> - Originally published in 1915, this important publication on Arizona Mining has not been available for nearly a century. Included are rare insights into hundreds of gold, silver, copper and other mines in this famous Arizona mining area. Details include the locations, geology, history, production and other facts of the mines of this region. **8.5" X 11", 394 ppgs. Retail Price: $24.99**

<u>Mining in the Bisbee Quadrangle of Arizona</u> - Originally published in 1906, this important publication on Arizona Mining has not been available for nearly a century. Included are rare insights into hundreds of gold, silver, copper and other mines in this famous Arizona mining area. Details include the locations, geology, history, production and other facts of the mines of this important mining region. **8.5" X 11", 188 ppgs. Retail Price: $14.99**

<u>Placer Gold Mining in Arizona</u> - Unavailable since 1922, this hard to find publication includes valuable insights into the placer gold mines of the Arizona. Originally released as "Placer Gold of Arizona", despite its small size, this publication includes important details into the history, geology and locations of the well known placer gold mines and prospects in the State of Arizona. **48 ppgs, 8.99**

<u>Gold and Copper Mining near Payson, Arizona</u> - Written in 1915, this hard to find publication includes valuable insights into the gold and copper mining industry of Arizona. Highlighted here are the gold and copper mines near Payson, Arizona. **68 ppgs, 8.99**

<u>Lode Gold Mining in Arizona</u> - Unavailable since 1934, this hard to find publication, originally released as "Arizona Lode Gold Mines and Gold Mining" includes valuable insights into the gold mining industry of Arizona. Included are valuable insights into over 150 hardrock gold mines in over 30 different mining districts in Arizona. **278 ppgs, 21.99**

<u>Mining in the Dragoon Quadrangle of Cochise County, Arizona</u> - Unavailable since 1964, this hard to find publication includes valuable insights into the mines of the Dragoon Quadrangle Mining Region. The publication includes important details into the history, geology and locations of the well known mines and prospects in this famous mining region of Arizona. **224 ppgs., 19.99**

<u>Directory of Operating Mines in Arizona in 1915</u> - Unavailable since 1916, this hard to find publication includes valuable insights into the mines of Arizona. This small publication includes a complete list of the mines that were operating in the State of Arizona during 1915 and includes details such as general location, owners and some basic facts about each mining operation. **52 ppgs. 8.99**

<u>Arizona Ore Deposits</u> - Unavailable since 1938, this hard to find publication includes valuable insights into some ore deposits of Arizona. Included are valuable insights into the formation and characteristics of valuable ore deposits in the Jerome, Miami, Inspiration, Clifton, Morenci, Ray, Ajo, Eureka, Tombstone and Magma mining districts. Included are details into some of the major gold, silver and copper mines of these important Arizona mining areas. **160 ppgs, 14.99**

Montana Mining Books

A History of Butte Montana: The World's Greatest Mining Camp - First published in 1900 by H.C. Freeman, this important publication sheds a bright light on one of the most important mining areas in the history of The West. Together with his insights, as well as rare photographs of the periods, Harry Freeman describes Butte and its vicinity from its early beginnings, right up to its flush years when copper flowed from its mines like a river. At the time of publication, Butte, Montana was known worldwide as "The Richest Mining Spot On Earth" and produced not only vast amounts of copper, but also silver, gold and other metals from its mines. Freeman illustrates, with great detail, the most important mines in the vicinity of Butte, providing rare details on their owners, their history and most importantly, how the mines operated and how their treasures were extracted. Of particular interest are the dozens of rare photographs that depict mines such as the famous Anaconda, the Silver Bow, the Smoke House, Moose, Paulin, Buffalo, Little Minah, the Mountain Consolidated, West Greyrock, Cora, the Green Mountain, Diamond, Bell, Parnell, the Neversweat, Nipper, Original and many others. 8.5" X 11", 142 ppgs. Retail Price: $12.99

The Butte Mining District of Montana - This important publication on Montana Mining has not been available for over a century. Included are rare insights into the gold, copper and silver mines of Butte, Montana together with hard to find maps and photographs. Some of the topics include the early history of gold, silver and copper mining in the Butte area, insight into the geology of its mining areas, the local distribution of gold, silver and copper ores, as well their composition and how to identify them. Also included are detailed facts about the mines in the Butte Mining District, including the famous Anaconda Mine, Gagnon, Parrot, Blue Vein, Moscow, Poulin, Stella, Buffalo, Green Mountain, Wake Up Jim, the Diamond-Bell Group, Mountain Consolidated, East Greyrock, West Greyrock, Snowball, Corra, Speculator, Adirondack, Miners Union, the Jessie-Edith May Group, Otisco, Iduna, Colorado, Lizzie, Cambers, Anderson, Hesperus, Preferencia and dozens of others. 8.5" X 11", 298 ppgs. Retail Price: $24.99

Mines of the Helena Mining Region of Montana - This important publication on Montana Mining has not been available for over a century. Included are rare insights into the gold, copper and silver mines of the vicinity of Helena, Montana, including the Marysville Mining District, Elliston Mining District, Rimini Mining District, Helena Mining District, Clancy Mining District, Wickes Mining District, Boulder and Basin Mining Districts and the Elkhorn Mining District. Some of the topics include the early history of gold, silver and copper mining in the Helena area, insight into the geology of its mining areas, the local distribution of gold, silver and copper ores, as well their composition and how to identify them. Also included are detailed facts, history, geology and locations of over one hundred gold, silver and copper mines in the area . 8.5" X 11", 162 ppgs, Retail Price: $14.99

Mines and Geology of the Garnet Range of Montana - This important publication on Montana Mining has not been available for over a century. Included are rare insights into the gold, copper and silver mines of the vicinity of this important mining area of Montana. Some of the topics include the early history of gold, silver and copper mining in the Garnet Mountains, insight into the geology of its mining areas, the local distribution of gold, silver and copper ores, as well their composition and how to identify them. Also included are detailed facts, history, geology and locations of numerous gold, silver and copper mines in the area . 8.5" X 11", 100 ppgs, Retail Price: $11.99

Mines and Geology of the Philipsburg Quadrangle of Montana - This important publication on Montana Mining has not been available for over a century. Included are rare insights into the gold, copper and silver mines of the vicinity of this important mining area of Montana. Some of the topics include the early history of gold, silver and copper mining in the Philipsburg Quadrangle, insight into the geology of its mining areas, the local distribution of gold, silver and copper ores, as well their composition and how to identify them. Also included are detailed facts, history, geology and locations of over one hundred gold, silver and copper mines in the area 8.5" X 11", 290 ppgs, Retail Price: $24.99

Geology of the Marysville Mining District of Montana - Included are rare insights into the mining geology of the Marysville Mining District. Some of the topics include the early history of gold, silver and copper mining in the area, insight into the geology of its mining areas, the local distribution of gold, silver and copper ores, as well their composition and how to identify them. Also included are detailed facts, history, geology and locations of gold, silver and copper mines in the area 8.5" X 11", 198 ppgs, Retail Price: $19.99

The Geology and Mines of Northern Idaho and North Western Montana- See listing under Idaho.

The History of Gold Dredging in Montana - Unavailable since 1916, this important publication was originally published by the Us Bureau of Mines and has been unavailable for a century. A century and more ago, giant dredging machines dug in Montana's rivers and creeks in search of illusive golden riches. First appearing in California in the 1850's, gold dredges finally reached their peak of development in Siberia and New Zealand before becoming popular again in the United States. This book offers a unique historical perspective on the gold dredges that once operated in Montana. This book on Montana mining history is lavishly illustrated with dozens of rare historic photos gold dredges that once operated in Montana, as well as hard to locate plans on how these dredges were designed. 120 ppgs., 11.99

Nevada Mining Books

The Bull Frog Mining District of Nevada - Unavailable since 1910, this publication was originally compiled by the United States Department of Interior. This volume also includes important insights into the geologic formations, faults and other aspects of economic geology in this Nevada mining district. Of particular interest are the fine details on many mines in the area, including their locations, histories, development and mineralization. Some of the mines featured include the National Bank Mine, Providence, Gibraltor, Tramps, Denver, Original Bullfrog, Gold Bar, Mayflower, Homestake-King and other mines and prospects. 8.5" X 11", 152 ppgs, Retail Price: $14.99

History of the Comstock Lode - Unavailable since 1876, this publication was originally released by John Wiley & Sons. This volume also includes important insights into the famous Comstock Lode of Nevada that represented the first major silver discovery in the United States. During its spectacular run, the Comstock produced over 192 million ounces of silver and 8.2 million ounces of gold. Not only did the Comstock result in one of the largest mining rushes in history and yield immense fortunes for its owners, but it made important contributions to the development of the State of Nevada, as well as neighboring California. Included here are important details on not only the early development and history of the Comstock, but also rare early insight into its mines, ore and its geology.8.5" X 11", 244 ppgs, Retail Price: $19.99

The Pioche Mining District of Nevada - First published in 1932, it has been unavailable for over a century and sheds important light on the mining history of Nevada. Some of the topics include the history of mining in this district, as well as the characteristics of its mineral and ore deposits. Also included are insights into the history, production, characteristics and locations of numerous mines in the area. Some of the mines include the Combined Metals, Pioche, Ely Valley, No. 10, Poorman, Wide Awake, Alps, Prince, Virginia Louise, Half Moon, Abe Lincoln, Fairview, Bristol Silver, National, Vesuvius, Inman, Tempest, Hillside, Jackrabbit, Lucky Star, Fortuna, Mendha, Manhattan, Hamburg, Comet, Lyndon and others. 108 ppgs 10.99

The Yerington Mining District of Nevada - First published in 1932, it has been unavailable for over a century and sheds important light on the mining history of Nevada. Some of the topics include the history of mining in this district, as well as the characteristics of its mineral and ore deposits. Also included are insights into the history, production, characteristics and locations of numerous mines in the area. Some of the mines include the Bluestone, Mason Valley, Malachite, McConnell, Greenwood, Western Nevada, Ludwig, Douglas Hill, Casting Copper, Montana-Yerington, Empire, Jim Beatty, Terry and McFarland, Blue Jay and others. 92 ppgs, 10.99

The Genesis of the Ores of Tonopah Nevada - Unavailable since 1918, this hard to find publication includes valuable insights into the gold mines around Tonopah, Nevada. The publication includes important details into the geology of mines in the Tonopah Mining District of Nevada. 90 ppgs, 10.99

Mining Camps of Elko, Lander and Eureka Counties Nevada - Unavailable since 1910, this hard to find publication includes valuable insights into the mining camps of Elko, Lander and Eureka Counties, Nevada. The publication includes important details into the history of mines and mining in these three Nevada counties. 154 ppgs, 12.99

Ore Deposits of the Bullfrog Quadrangle - Unavailable since 1964 and released as "Geology of Bullfrog Quadrangle and Ore Deposits Related to Bullfrog Hills Caldera, Nye County, Nevada and Inyo County, California". The publication includes important details into the geology of mines in the Bullfrog Quadrangle of Nye County, Nevada and Inyo County, California. 52 ppgs, 9.99

Mining in Eureka County Nevada - Unavailable since 1879, this hard to find publication includes valuable insights into the early mining history off Eureka County, Nevada. The publication includes important details into the early history of the mines of Eureka County, as well as their development, production and how their ores were treated. Also included are details on the 1872 Mining Act, as well as the local rules, regulations and customs of the miners in Eureka County.134 ppgs, 12.99

Colorado Mining Books

Ores of The Leadville Mining District - Unavailable since 1926, this publication was originally compiled by the United States Department of Interior. This volume also includes important insights into the ores and mineralization of the Leadville Mining District in Colorado. Topics include historic ore prospecting methods, local geology, insights into ore veins and stockworks, the local trend and distribution of ore channels, reverse faults, shattered rock above replacement ore bodies, mineral enrichment in oxidized and sulphide zones and more. **8.5" X 11", 66 ppgs, Retail Price: $8.99**

Mining in Colorado - Unavailable since 1926, this publication was originally compiled by the United States Department of Interior. This volume also includes important insights into the mining history of Colorado from its early beginnings in the 1850's right up to the mid 1920's. Not only is Colorado's gold mining heritage included, but also its silver, copper, lead and zinc mining industry. Each mining area is treated separately, detailing the development of Colorado's mines on a county by county basis. **8.5" X 11", 284 ppgs, Retail Price: $19.99**

Gold Mining in Gilpin County Colorado - Unavailable since 1876, this publication was originally compiled by the Register Steam Printing House of Central City, Colorado. A rare glimpse at the gold mining history and early mines of Gilpin County, Colorado from their first discovery in the 1850's up to the "flush years" of the mid 1870's. Of particular interest is the history of the discovery of gold in Gilpin County and details about the men who made those first strikes. Special focus is given to the early gold mines and first mining districts of the area, many of which are not detailed in other books on Colorado's gold mining history. **8.5" X 11", 156 ppgs, Retail Price: $12.99**

Mining in the Gold Brick Mining District of Colorado - Important insights into the history of the Gold Brick Mining District, as well as its local geography and economic geology. Also included are the histories and locations of historic mines in this important Colorado Mining District, including the Cortland, Carter, Raymond, Gold Links, Sacramento, Bassick, Sandy Hook, Chronicle, Grand Prize, Chloride, Granite Mountain, Lucille, Gray Mountain, Hilltop, Maggie Mitchell, Silver Islet, Revenue, Roosevelt, Carbonate King and others. In addition to hardrock mining, are also included are details on gold placer mining in this portion of Colorado. **8.5" X 11", 140 ppgs, Retail Price: $12.99**

Ore Deposits of the London Fault of Colorado - First published in 1941, it has been unavailable since those days and sheds important light on the mines and mineral deposits of the London Fault in Central Colorado's Alma Mining District. This publication sheds important light on the gold veins and lead-silver deposits of the Alma Mining District. Included are geologic details on the London Mine, American Mine, Havigorst Tunnel, Ophir Mine, Mosher Tunnel, London-Butte Mine, Venture Shaft, Hard-To-Beat Mine, Oliver Twist Tunnel, Sacramento Mine, Mudsill Mine, Sherwood Mine, Wagner, Barcoe Tunnel and other mines in this important mining region. 110 ppgs., 10.99

The Mines of Colorado - First published in 1867, it has been unavailable since those days and sheds important light on Colorado's early mining history. Written shortly after the events took place, this publication sheds important light on the Pike's Peak Gold Rush, the discovery of gold on Ralston Creek and Dry Creek in the 1850's, as well as details on the first wave of miners into Colorado and their trials and tribulations as they crossed the Great Plains. Also included are details on early discoveries of lode gold in the mountainous regions of Colorado, details on the early mines hardrock and placer mines, and much more. It is a veritable treasure trove on Colorado's early mining history and will be of great importance to anyone who is interested in the mining of gold or other minerals in Colorado, as well as those interested in the history of the state. 478 ppgs., 29.99

The La Plata Mining District of Colorado - Originally titled "Geology and Ore Deposits in the Vicinity of the La Plata District of Colorado" and first published in 1949, it has been unavailable since those days and sheds important light on the mines and mineral deposits of the La Plata Mining District of Colorado. 214 ppgs., 19.99

Washington Mining Books

The Republic Mining District of Washington - Unavailable since 1910, this important publication was originally published by the Washington Geologic Survey and has been unavailable for a century. Topics include the geology, rock formations and the formation of ore deposits in this important mining area of Washington State. Also included are hard to find details on the geology, history and locations of dozens of mines in the area. Some of the mines featured include the New Republic Mine, Ben Hur, Morning Glory, the South Republic Mine, Quilp, Surprise, Black Tail, Lone Pine, San Poil, Mountain Lion, Tom Thumb, Elcaliph and many others. **8.5" X 11", 94 ppgs, Retail Price: $10.99**

The Myers Creek and Nighthawk Mining Districts of Washington - Unavailable since 1911, this important publication was originally published by the Washington Geologic Survey and has been unavailable for a century. Topics include the geology, rock formations and the formation of ore deposits in these important mining areas of Washington State. Also included are hard to find details on the geology, history and locations of dozens of mines in the area. Some of the mines featured include the Grant Mine, Monterey, Nip and Tuck, Myers Creek, Number Nine, Neutral, Rainbow, Aztec, Crystal Butte, Apex, Butcher Boy, Molson, Mad River, Olentangy, Delate, Kelsey, Golden Chariot, Okanogan, Ohio, Forty-Ninth Parallel, Nighthawk, Favorite, Little Chopaka, Summit, Number One, California, Peerless, Caaba, Prize Group, Ruby, Mountain Sheep, Golden Zone, Rich Bar, Similkameen, Kimberly, Triune, Hiawatha, Trinity, Hornsilver, Maquae, Bellevue, Bullfrog, Palmer Lake, Ivanhoe, Copper World and many others. **8.5" X 11", 136 ppgs, Retail Price: $12.99**

The Blewett Mining District of Washington - Unavailable since 1911, this important publication was originally published by the Washington Geologic Survey and has been unavailable for a century. Topics include the geology, rock formations and the formation of ore deposits in this important mining area of Washington State. Also included are hard to find details on the geology, history and locations of dozens of mines in the area. Some of the mines featured include the Washington Meteor, Alta Vista, Pole Pick, Blinn, North Star, Golden Eagle, Tip Top, Wilder, Golden Guinea, Lucky Queen, Blue Bell, Prospect, Homestake, Lone Rock, Johnson, and others. **8.5" X 11", 134 ppgs, Retail Price: $12.99**

Silver Mining In Washington - Unavailable since 1955, this important publication was originally published by the Washington Geologic Survey. Featured are the hard to find locations and details pertaining to Washington's silver mines. **8.5" X 11", 180 ppgs, Retail Price: $15.99**

The Mines of Snohomish County Washington - Unavailable since 1942, this important publication was originally published by the Washington Geologic Survey and has been unavailable for seventy years. Featured are details on a large number of gold, silver, copper, lead and other metallic mineral mines. Included are the locations of each historic mine, along with information on the commodity produced. **8.5" X 11", 98 ppgs, Retail Price: $10.99**

The Mines of Chelan County Washington - Unavailable since 1943, this important publication was originally published by the Washington Geologic Survey and has been unavailable for seventy years. Featured are details on a large number of gold, silver, copper, lead and other metallic mineral mines. Included are the locations of each historic mine, along with information on the commodity. **8.5" X 11", 88 ppgs, Retail Price: $9.99**

Metal Mines of Washington - Unavailable since 1921, this important publication was originally published by the Washington Geologic Survey and has been unavailable for nearly ninety years. Widely considered a masterpiece on the Washington Mining Industry, "Metal Mines of Washington" sheds light on the important details of Washington's early mining years. Featured are details on hundreds of gold, silver, copper, lead and other metallic mineral mines. Included are hard to find details on the mineral resources of this state, as well as the locations of historic mines. Lavishly illustrated with maps and historic photos and complete with a glossary to explain any technical terms found in the text, this is one of the most important works on mining in the State of Washington. No prospector or miner should be without it if they are interested in mining in Washington. **8.5" X 11", 396 ppgs, Retail Price: $24.99**

Gem Stones In Washington - Unavailable since 1949, this important publication was originally published by the Washington Geologic Survey and has been unavailable since first published. Included are details on where to find naturally occurring gem stones in the State of Washington, including quartz crystal, amethyst, smoky quartz, milky quartz, agates, bloodstone, carnelian, chert, flint, jasper, onyx, petrified wood, opal, fire opal, hyalite and others. **8.5" X 11", 54 ppgs, Retail Price: $8.99**

The Covada Mining District of Washington - Unavailable since 1913, this important publication was originally published by the Washington Geologic Survey and has been unavailable for a century. Topics include the geology, rock formations and the formation of ore deposits in this important mining area of Washington State. Also included are hard to find details on the geology, history and locations of dozens of mines in the area. Some of the mines featured include the Admiral, Advance, Algonkian, Big Bug, Big Chief, Big Joker, Black Hawk, Black Tail, Black Thorn, Captain, Cherokee Strip, Colorado, Dan Patch, Dead Shot, Etta, Good Ore, Greasy Run, Great Scott, Idora, IXL, Jay Bird, Kentucky Bell, King Solomon, Laurel, Laura S, Little Jay, Meteor, Neglected, Northern Light, Old Nell, Plymouth Rock, Polaris, Quandary, Reserve, Shoo Fly, Silver Plume, Three Pines, Vernie, White Rose and dozens of others. **8.5" X 11", 114 ppgs, Retail Price: $10.99**

The Index Mining District of Washington - Unavailable since 1912, this important publication was originally published by the Washington Geologic Survey and has been unavailable for a century. Topics include the geology, rock formations and the formation of ore deposits in this important mining area of Washington State. Also included are hard to find details on the geology, history and locations of dozens of mines in the area. Some of the mines featured include the Sunset, Non-Pareil, Ethel Consolidated, Kittaning, Merchant, Homestead, Co-operative, Lost Creek, Uncle Sam, Calumet, Florence-Rae, Bitter Creek, Index Peacock, Gunn Peak, Helena, North Star, Buckeye. Copper Bell, Red Cross and others. 8.5" X 11", 114 ppgs, Retail Price: $11.99

Mining & Mineral Resources of Stevens County Washington - Unavailable since 1920, this important publication was originally published by the Washington Geologic Survey and has been unavailable for a century. Topics include the geology, rock formations and the formation of ore deposits in these important mining areas of Washington State. Also included are hard to find details on the geology, history and locations of hundreds of mines in the area. 8.5" X 11", 372 ppgs, Retail Price: $24.99

The Mines and Geology of the Loomis Quadrangle Okanogan County, Washington - Unavailable since 1972, this important publication was originally published by the Washington Geologic Survey and has been unavailable for a century. Topics include the geology, rock formations and the formation of ore deposits in this important mining area of Washington State. Also included are hard to find details on the geology, history and locations of dozens of gold, copper, silver and other mines in the area. 8.5" X 11", 150 ppgs, Retail Price: $12.99

The Conconully Mining District of Okanogan County Washington - Unavailable since 1973, this important publication was originally published by the Washington Geologic Survey and has been unavailable for a century. Topics include the geology, rock formations and the formation of ore deposits in this important mining area of Washington State, which also includes Salmon Creek, Blue Lake and Galena. Also included are hard to find details on the geology, mining history and locations of dozens of mines in the area. Some of the mines include Arlington, Fourth of July, Sonny Boy, First Thought, Last Chance, War Eagle-Peacock, Wheeler, Mohawk, Lone Star, Woo Loo Moo Loo, Keystone, Hughes, Plant-Callahan, Johnny Boy, Leuena, Gubser, John Arthur, Tough Nut, Homestake, Key and many others 8.5" X 11", 68 ppgs, Retail Price: $8.99

Wyoming Mining Books

Mining in the Laramie Basin of Wyoming - Unavailable since 1909, this publication was originally compiled by the United States Department of Interior. Also included are insights into the mineralization and other characteristics of this important mining region, especially in regards to coal, limestone, gypsum, bentonite clay, cement, sand, clay and copper. 8.5" X 11", 104 ppgs, Retail Price: $11.99

New Mexico Mining Books

The Mogollon Mining District of New Mexico - Unavailable since 1927, this important publication was originally published by the US Department of Interior and has been unavailable for 80 years. Topics include the geology, rock formations and the formation of ore deposits in this important mining area in New Mexico. Of particular focus is information on the history and production of the ore deposits in this area, their form and structure, vein filling, their paragenesis, origins and ore shoots, as well as oxidation and supergene enrichment. Also included are hard to find details, including the descriptions and locations of numerous gold, silver and other types of mines, including the Eureka, Pacific, South Alpine, Great Western, Enterprise, Buffalo, Mountain View, Floride, Gold Dust, Last Chance, Deadwood, Confidence, Maud S., Deep Down, Little Fanney, Trilby, Johnson, Alberta, Comet, Golden Eagle, Cooney, Queen, the Iron Crown, Eberle, Clifton, Andrew Jackson mine, Mascot and others. 8.5" X 11", 144 ppgs, Retail Price: $12.99

The Percha Mining District of Kingston New Mexico - Unavailable since 1883, this important publication was originally published by the Kingston Tribune and has been unavailable for over one hundred and thirty five years. Having been written during the earliest years of gold and silver mining in the Percha Mining District, unlike other books on the subject, this work offers the unique perspective of having actually been written while the early mining history of this area was still being made. In fact, the work was written so early in the development of this area that many of the notable mines in the Percha District were less than a few years old and were still being operated by their original discoverers with the same enthusiasm as when they were first located. Included are hard to find details on the very earliest gold and silver mines of this important mining district near Kingston in Sierra County, New Mexico. 8.5" X 11", 68 ppgs, Retail Price: $9.99

East Coast Mining Books

<u>The Gold Fields of the Southern Appalachians</u> - Unavailable since 1895, this important publication was originally published by the US Department of Interior and has been unavailable for nearly 120 years. Topics include the geology, rock formations and the formation of ore deposits in this important mining area of the American South. Of particular focus is information on the history and statistics of the ore deposits in this area, their form and structure and veins. Also included are details on the placer gold deposits of the region. The gold fields of the Georgian Belt, Carolinian Belt and the South Mountain Mining District of North Carolina are all treated in descriptive detail. Included are hard to find details, including the descriptions and locations of numerous gold mines in Georgia, North Carolina and elsewhere in the American South. Also included are details on the gold belts of the British Maritime Provinces and the Green Mountains. **8.5" X 11", 104 ppgs, Retail Price: $9.99**

Gold Rush Tales Series

<u>Millions in Siskiyou County Gold</u> - In this first volume of the "Gold Rush Tales" series, leading mining historian and editor Kerby Jackson, introduces us to the story of how millions of dollars worth of gold was discovered in Siskiyou County during the California Gold Rush. Lavishly illustrated with photos from the 19th Century, this hard to find information was first published in 1897 and sheds important light onto the gold rush era in Siskiyou County, California and the experiences of the men who dug for the gold and actually found it. **8.5" X 11", 82 ppgs, Retail Price: $9.99**

<u>The California Rand in the Days of '49</u> - In this second volume of the "Gold Rush Tales" series, leading mining historian and editor Kerby Jackson, introduces us to four tales from the California Gold Rush. Lavishly illustrated with photos from the 19th Century, this hard to find information was first published in 1890's and includes the stories of "California's Rand", details about Chinese miners, how one early miner named Baker struck it rich and also the story of Alphonzo Bowers, who invented the first hydraulic gold dredge. **8.5" X 11", 54 ppgs, Retail Price: $9.99**

More Mining Books

<u>Prospecting and Developing A Small Mine</u> - Topics covered include the classification of varying ores, how to take a proper ore sample, the proper reduction of ore samples, alluvial sampling, how to understand geology as it is applied to prospecting and mining, prospecting procedures, methods of ore treatment, the application of drilling and blasting in a small mine and other topics that the small scale miner will find of benefit. **8.5" X 11", 112 ppgs, Retail Price: $11.99**

<u>Timbering For Small Underground Mines</u> - Topics covered include the selection of caps and posts, the treatment of mine timbers, how to install mine timbers, repairing damaged timbers, use of drift supports, headboards, squeeze sets, ore chute construction, mine cribbing, square set timbering methods, the use of steel and concrete sets and other topics that the small underground miner will find of benefit. This volume also includes twenty eight illustrations depicting the proper construction of mine timbering and support systems that greatly enhance the practical usability of the information contained in this small book. **8.5" X 11", 88 ppgs. Retail Price: $10.99**

<u>Timbering and Mining</u> - A classic mining publication on Hard Rock Mining by W.H. Storms. Unavailable since 1909, this rare publication provides an in depth look at American methods of underground mine timbering and mining methods. Topics include the selection and preservation of mine timbers, drifting and drift sets, driving in running ground, structural steel in mine workings, timbering drifts in gravel mines, timbering methods for driving shafts, positioning drill holes in shafts, timbering stations at shafts, drainage, mining large ore bodies by means of open cuts or by the "Glory Hole" system, stoping out ore in flat or low lying veins, use of the "Caving System", stoping in swelling ground, how to stope out large ore bodies, Square Set timbering on the Comstock and its modifications by California miners, the construction of ore chutes, stoping ore bodies by use of the "Block System", how to work dangerous ground, information on the "Delprat System" of stoping without mine timbers, construction and use of headframes and much more. This volume provides a reference into not only practical methods of mining and timbering that may be employed in narrow vein mining by small miners today, but also rare insights into how mines were being worked at the turn of the 19th Century. **8.5" X 11", 288 ppgs. Retail Price: $24.99**

A Study of Ore Deposits For The Practical Miner - Mining historian Kerby Jackson introduces us to a classic mining publication on ore deposits by J.P. Wallace. First published in 1908, it has been unavailable for over a century. Included are important insights into the properties of minerals and their identification, on the occurrence and origin of gold, on gold alloys, insights into gold bearing sulfides such as pyrites and arsenopyrites, on gold bearing vanadium, gold and silver tellurides, lead and mercury tellurides, on silver ores, platinum and iridium, mercury ores, copper ores, lead ores, zinc ores, iron ores, chromium ores, manganese ores, nickel ores, tin ores, tungsten ores and others. Also included are facts regarding rock forming minerals, their composition and occurrences, on igneous, sedimentary, metamorphic and intrusive rocks, as well as how they are geologically disturbed by dikes, flows and faults, as well as the effects of these geologic actions and why they are important to the miner. Written specifically with the common miner and prospector in mind, the book will help to unlock the earth's hidden wealth for you and is written in a simple and concise language that anyone can understand. **8.5" X 11", 366 ppgs. Retail Price: $24.99**

Mine Drainage - Unavailable since 1896, this rare publication provides an in depth look at American methods of underground mine drainage and mining pump systems. This volume provides a reference into not only practical methods of mining drainage that may be employed in narrow vein mining by small miners today, but also rare insights into how mines were being worked at the turn of the 19th Century. **8.5" X 11", 218 ppgs. Retail Price: $24.99**

Fire Assaying Gold, Silver and Lead Ores - Unavailable since 1907, this important publication was originally published by the Mining and Scientific Press and was designed to introduce miners and prospectors of gold, silver and lead to the art of fire assaying. Topics include the fire assaying of ores and products containing gold, silver and lead; the sampling and preparation of ore for an assay; care of the assay office, assay furnaces; crucibles and scorifiers; assay balances; metallic ores; scorification assays; cupelling; parting' crucible assays, the roasting of ores and more. This classic provides a time honored method of assaying put forward in a clear, concise and easy to understand language that will make it a benefit to even beginners. **8.5" X 11", 96 ppgs. Retail Price: $11.99**

Methods of Mine Timbering - Originally published in 1896, this important publication on mining engineering has not been available for nearly a century. Included are rare insights into historical methods of timbering structural support that were used in underground metal mines during the California that still have a practical application for the small scale hardrock miner of today. **8.5" X 11", 94 ppgs. Retail Price: $10.99**

The Enrichment of Copper Sulfide Ores - First published in 1913, it has been unavailable for over a century. Topics include the definition and types of ore enrichment, the oxidation of copper ores, the precipitation of metallic sulfides. Also included are the results of dozens of lab experiments pertaining to the enrichment of sulfide ores that will be of interest to the practical hard rock mine operator in his efforts to release the metallic bounty from his mine's ore. **8.5" X 11", 92 ppgs. Retail Price: $9.99**

A Study of Magmatic Sulfide Ores - Unavailable since 1914, this rare publication provides an in depth look at magmatic sulfide ores. Some of the topics included are the definition and classification of magmatic ores, descriptions of some magmatic sulfide ore deposits known at the time of publication including copper and nickel bearing pyrrohitic ore bodies, chalcopyrite-bornite deposits, pyritic deposits, magnetite-ileminite deposits, chromite deposits and magmatic iron ore deposits. Also included are details on how to recognize these types of ore deposits while prospecting for valuable hardrock minerals. **8.5" X 11", 138 ppgs. Retail Price: $11.99**

The Cyanide Process of Gold Recovery - Unavailable since 1894 and released under the name "The Cyanide Process: Its Practical Application and Economical Results", this rare publication provides an in depth look at the early use of cyanide leaching for gold recovery from hardrock mine ores. This volume provides a reference into the early development and use of cyanide leaching to recover gold. **8.5" X 11", 162 ppgs. Retail Price: $14.99**

California Gold Milling Practices - Unavailable since 1895 and released under the name "California Gold Practices", this rare publication provides an in depth look at early methods of milling used to reduce gold ores in California during the late 19th century. This volume provides a reference into the early development and use of milling equipment during the earliest years of the California Gold Rush up to the age of the Industrial Revolution. Much of the information still applies today and will be of use to small scale miners engaging in hardrock mining. **8.5" X 11", 104 ppgs. Retail Price: $10.99**

Leaching Gold and Silver Ores With The Plattner and Kiss Processes - Mining historian Kerby Jackson introduces us to a classic mining publication on the evaluation and examination of mines and prospects by C.H. Aaron. First published in 1881, it has been unavailable for over a century and sheds important light on the leaching of gold and silver ores with the Plattner and Kiss processes. **8.5" X 11", 204 ppgs. Retail Price: $15.99**

The Metallurgy of Lead and the Desilverization of Base Bullion - First published in 1896, it has been unavailable for over a century and sheds important light on the the recovery of silver from lead based ores. Some of the topics include the properties of lead and some of its compounds, lead ores such as galenite, anglesite, cerussite and others, the distribution of lead ores throughout the United States and the sampling and assaying of lead ores. Also covered is the metallurgical treatment of lead ores, as well as the desilverization of lead by the Pattinson Process and the Parkes Process. Hofman's text has long been considered one of the most important early works on the recovery of silver from lead based ores. 8.5" X 11", 452 ppgs. Retail Price: $29.99

Ore Sampling For Small Scale Miners - First published in 1916, it has been unavailable for over a century and sheds important light on historic methods of ore sampling in hardrock mines. Topics include how to take correct ore samples and the conditions that affect sampling, such as their subdivision and uniformity. Particular detail is given to methods of hand sampling ore bodies by grab sample, pipe sample and coning, as well as sampling by mechanical methods. Also given are insights into the screening, drying and grinding processes to achieve the most consistent sample results and much more. 8.5" X 11", 124 ppgs. Retail Price: $12.99

The Extraction of Silver, Copper and Tin from Ores - First published in 1896, it has been unavailable for over a century and sheds important light on how historic miners recovered silver, copper and tin from their mining operations. The book is split into three sections, including a discussion on the Lixiviation of Silver Ores, the mining and treatment of copper ores as practiced at Tharsis, Spain and the smelting of tin as it was practiced by metallurgists at Pulo Brani, Singapore. Also included is an overview and analysis of these historic metal recovery methods that will be of benefit to those interested in the extraction of silver, copper and tin from small mines. 8.5" X 11", 118 ppgs. Retail Price: $14.99

The Roasting of Gold and Silver Ores - First published in 1880, it has been unavailable for over a century and sheds important light on how historic miners recovered gold and silver rom their mining operations. Topics include details on the most important silver and free milling gold ores, methods of desulphurization of ores, methods of deoxidation, the chlorination of ores, methods and details on roasting gold and silver ores, notes on furnaces and more. Also included are details on numerous methods of gold and silver recovery, including the Ottokar Hofman's Process, the Patera Process, Kiss Process, Augustin Process, Ziervogel Process and others. 8.5" X 11", 178 ppgs. Retail Price: $19.99

The Examination of Mines and Prospects - First published in 1912, it has been unavailable for over a century and sheds important light on how to examine and evaluate hardrock mines, prospects and lode mining claims. Sections include Mining Examinations, Structural Geology, Structural Features of Ore Deposits, Primary Ores and their Distribution, Types of Primary Ore Deposits, Primary Ore Shoots, The Primary Alteration of Wall Rocks, Alterations by Surface Agencies, Residual Ores and their Distribution, Secondary Ores and Ore Shoots and Vein Outcrops. This hard to find information is a must for those who are interested in owning a mine or who already own a lode mining claim and wish to succeed at quartz mining. 8.5" X 11", 250 ppgs. Retail Price: $19.99

Garnets: Their Mining, Milling and Utilization - First published in 1925, it has been unavailable since those days and sheds important light on the mining, milling and utilization of garnets. Included are details on the characteristics of garnets, where they are found and how they were mined. 78 ppgs, 10.99

Gemstones and Precious Stones of North America - Leading mining historian Kerby Jackson introduces us to a classic mining publication on the gems and precious stones of the United States, Canada and mexico. First published in 1890, it has been unavailable since those days and sheds important light on the gems and precious stones that may be found in North America. Included are chapters on diamonds, corundum, sapphire, ruby, topaz, emerald, disapore, spinel, turquoise, tourmaline, garnets, beyrl, peridot, zircon, quartz crystals, feldspars, pearls and many others. Included are details on where these gems and precious stones may be found throughout North America, as well as their characteristics. 360 ppgs, 24.99

Mining Camps and Mining Districts - First released in 1885 by Charles Howard Shinn under the title "Mining Camps: A Study in American Frontier Government", this publication offers a unique look at how early gold miners established their own forms of representative government during the California Gold Rush. Drawing on the the early mining codes of mideviel German miners in the Harz Mountains, on the mining customs of the Cornish tin miners and early Spanish mining laws introduced into California, the miners established the first governments in the American West. 340 ppgs, 24.99

BLM Field Handbook for Mineral Examiners - Leading mining historian Kerby Jackson introduces us to a classic mining publication on mine evaluation. First published in 1962, this work sheds important light on the techniques of BLM Mineral Examiners to perform validity on mining claims. 132 ppgs, 10.99

<u>Six Months In The Gold Mines During The California Gold Rush</u> - Unavailable since 1850, this important work is a first hand account of one "49'ers" personal experience during the great California Gold Rush, shedding important light on one of the most exciting periods in the history of not only California, but also the world. Compiled from journals written between 1847 and 1849 by E. Gould Buffum, a native of New York, "Six Months In The Gold Mines During The California Gold Rush" offers a rare look into the day to day lives of the people who came to California to work in her gold mines when the state was still a great frontier. **8.5" X 11", 290 ppgs. Retail Price: $19.99**

<u>The Discovery of Gold in Australia</u> - **First published in 1852, it has been unavailable since those days and sheds important light on Australia's gold mining history. Included are rare communications between British agents and the British Crown when gold was first discovered in Australia in 1851. This rare text contains hard to find details on Australia's first mining camps and Britain's early attempts to provide for the orderly regulation of gold mines in that part of the world. Also of interest are hard to find extracts of articles that appeared in the early colonial newspapers that did their best to report on Australia's gold rush as it took place.**
102 ppgs, 10.99

www.ingramcontent.com/pod-product-compliance
Lightning Source LLC
Chambersburg PA
CBHW080835180526
45168CB00006B/2692